XX International Scientific Conference "RESHETNEV READINGS" 2016

IOP Conference Series: Materials Science and Engineering
Volume 255

Krasnoyarsk, Russia
9 – 12 November 2016

ISBN: 978-1-5108-5032-3
ISSN: 1757-8981

Printed from e-media with permission by:

Curran Associates, Inc.
57 Morehouse Lane
Red Hook, NY 12571

CURRAN ASSOCIATES INC. proceedings.com

Some format issues inherent in the e-media version may also appear in this print version.

Printed by Curran Associates, Inc. (2017)

For permission requests, please contact the Institute of Physics
at the address below.

Institute of Physics
Dirac House, Temple Back
Bristol BS1 6BE UK

Phone: 44 1 17 929 7481
Fax: 44 1 17 920 0979

techtracking@iop.org

Additional copies of this publication are available from:

Curran Associates, Inc.
57 Morehouse Lane
Red Hook, NY 12571 USA
Phone: 845-758-0400
Fax: 845-758-2633
Email: curran@proceedings.com
Web: www.proceedings.com

Table of contents

Volume 255

XX International Scientific Conference Reshetnev Readings-2016
9–12 November 2016, Krasnoyarsk, Russian Federation

Accepted papers received: 8 October 2017
Published online: 23 October 2017

Preface

Papers

IOP Conf. Series: Materials Science and Engineering **255** (2017) 011001 doi:10.1088/1757-899X/255/1/011001

XX Anniversary International Scientific Conference "RESHETNEV READINGS"

The International Scientific Conference "RESHETNEV READINGS" is dedicated to the memory of Mikhail Reshetnev, an outstanding scientist, chief-constructor of space-rocket systems and communication satellites.

The current volume represents selected proceedings of the XX International Anniversary Scientific Conference "RESHETNEV READINGS" held on November 09 – 12, 2016.

Plenary sessions, round tables, forums are regularly attended by famous scientists, developers, designers representing the space technology sector, as well as professionals and experts in IT industry. A number of outstanding academic figures expressed their interest in the event of such a level including Jaures Alferov, Vice-president of the Russian Academy of Sciences (RAS), Academician of RAS, Nobel laureate, Dirk Bochar, General Secretary of the European Federation of National Engineering Associates (FEANI), Prof. Yuri Gulyaev, Academician of RAS, Member of the Presidium of RAS, President of the International Union of Scientific and Engineering Associations, Director of the Institute of Radio Engineering and Electronics of RAS, as well as rectors of the largest universities in Russia, chief executives of well-known research enterprises and representatives of big businesses.

We would like to thank our regular sponsors such as JSC "Reshetnev Information Satellite Systems", JSC "Krasnoyarsk Engineering Plant", Central Design Bureau "Geophysics", Krasnoyarsk Region Authorities. These enterprises and companies are leading ones in the aerospace branch. It is a great pleasure to cooperate and train specialists for them.

IOP Conf. Series: Materials Science and Engineering **255** (2017) 011002 doi:10.1088/1757-899X/255/1/011002

Peer review statement

All papers published in this volume of *IOP Conference Series: Materials Science and Engineering* have been peer reviewed through processes administered by the proceedings Editors. Reviews were conducted by expert referees to the professional and scientific standards expected of a proceedings journal published by IOP Publishing.

Investigation of the process rotary turning by multi-faceted cutters

N S Indakov[1], A S Binchurov[1], Y I Gordeev[1], V B Yasinski[1], D I Kiselyov[1], A A Lepeshev[1,2]

[1]Siberian Federal University, Kirensky str. 26, Krasnoyarsk, 660074 Russia
[2] Krasnoyarsk Scientific Center, Siberian Branch, Russian Academy of Sciences, 50/50, Akademgorodok, Krasnoyarsk, 660074, Russia

E-mail: mexanixs@mail.ru

Abstract. Effect of geometrical and technological parameters of method rotary turning by multifaceted cutters (RTMC) on the quality of machining was investigated. Using the numerical, analytical and experimental methods the effect of cutting conditions on roughness was determined. The resulting semi-empirical dependences allow to appoint the cutting conditions with predict roughness parameters Ra, Rz, $R_{\max}$.

1. Introduction

In the field of metal cutting of materials widely used such methods of turning as: the vertex cutting, peak less cutting, autorotation and forced rotary turning. Each of these methods has its rational application. However, these conventional machining methods have significant drawbacks in durability of the tool life, chips formation; the use of expensive MQL's cooling; the speed limitation, performance of machining and require the development of alternative and innovative methods of surface formation [1-3].

Rotary cutting reduces tool temperatures and the amount of tool wear owing interrupted turning, thermal diffusion into the tool body, and thermal dissipation into the air [4,5]. Tool rotation distributes the cutting heat and tool wear to the entire cutting edge with a constantly changing contact point, and consequently improves the tool life [6-8]. Tool-workpiece contact time, the length of the cutting edge, and cutting conditions have great influences on surface roughness, however, there has been few studies concerning these influences [6]. The kinematic parameters of the turning rotary process are the starting point for the selection of cutting conditions and tool geometry [8,9].

The RTMK method, developed at the Krasnoyarsk Polytechnic Institute of the Siberian Federal University, is one of the promising methods for processing such details as shafts from viscous, plastic and composite materials, as well as high-alloy, heat-resistant steel [8-10]. The method RTMC promising multi-faceted for extra-heavy-duty processing of difficult – to-cut materials, alloy, heat-resistant, hard-steels and alloys can be used in aerospace and transport engineering.

The objective of this study was to investigate effect of geometrical and technological parameters of RTMC on the quality of machining.

2. Results and Discussion

The parameters of chip geometry and workpiece roughness varies depend on the tool cutting edge (angles, tool geometry) and cutting conditions was determined by analytical and experimental methods. In the formation of the treated surface involves three movements: the main motion (rotation of the tool) V_r, linear motion or traverse f_n and additional moving of flow feed (workpiece rotation) f_{az}. The versatile cutter performs forced rotation around its axis with respect to the rotating speed V_r [8-10]. At the same time a tool is reported rectilinear motion along the axis of the workpiece longitudinal feed fn. Thus the tool rotation axis is set perpendicular to the axis of rotation of the workpiece and the center is on the axis line [10].

The rotary cutter is a multi-faceted plate of the cup shape with a bore hole and the cutting part, consisting from the N cutting blades [8-12]. Each blade comprises a front and rear surfaces which form between them a rectilinear cutting edge. Each blade cutting edge located in a plane perpendicular to the axis of rotation formed the multi-faceted cutters. The processing circuit of a multi-faceted rotary cutter shown in Fig. 1.

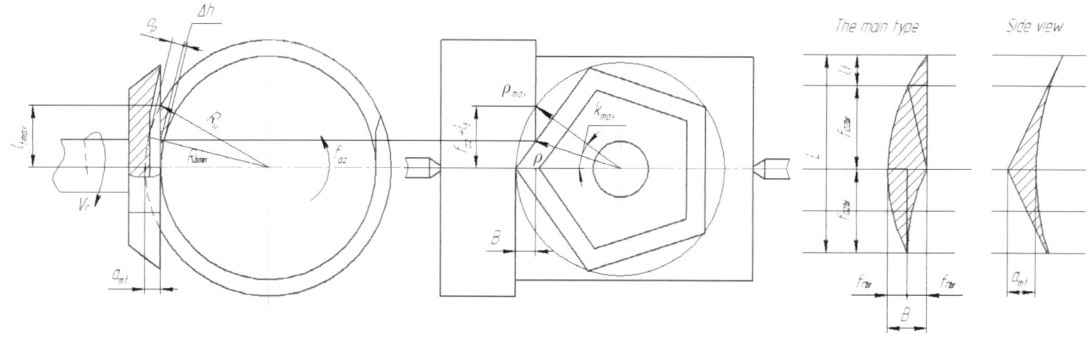

Figure 1. Kinematic scheme for determination the height of roughness (Δh)

The quality of the treated surface is estimated by the surface roughness and the state of the surface layer of the material. On Fig. 1 shows the estimated micro profile of cylindrical surface machined by RTMC. The height of surface irregularities determined by the formula:

$$\Delta h = \sqrt{\rho^2 \cdot \sin \kappa^2 + R_{\text{dmin}}^2} + R_{\text{dmin}}; \tag{1}$$

where: ρ - radius vector of the cutting edge contour, mm; $R_{d\min}$ – the radius of machined workpiece, mm; k – the angle between the center axis and the contact point of the cutting edge and the machined surface.

Longitudinal feed on the cutter face of f_{nbr} $=B/2$. Longitudinal cutter feed on the treated turnover shaft is calculated using the formula $f_n = f_{nbr} \cdot n_r \cdot N / n_b$ (n_r - turning number of a workpiece, turn/min; n_b - turning number of the tool, turn/min; N - number of cutter profile faces.); The azimuthal feed to the edge of the tool is $f_{azbr} = \rho_{max} \cdot sin\kappa_{max}$ (ρ_{max} - maximum radius vector of the cutting edge contour, mm; ρ_{min} - minimum radius vector of the cutting edge contour, mm). The number of revolutions of the workpiece $n_b = f_{azbr} \cdot n_r \cdot N/\pi \cdot d_b$. The number of revolutions of the tool is determined by $n_r = \pi \cdot d_3 \cdot n_b / f_{azbr} \cdot N$.

In order to avoid the kinematic undulation ($l_{t\max}$) is necessary to appoint azimuthal feed ($f_{az_{br}}$) using the equations:

$$l_{t\max} = \sqrt{R_W^2 - (R_W - a_{mt})^2}; \tag{2}$$

$$l_{t\max} \geq f_{az_{br}} + l_t \tag{3}$$

where: $l_{t\max}$ – height of exit the cutting blade from the cutting area under the plane of the centers, mm; R_W – the radius workpiece, mm.

The length (l_t) can be determined by the formula:

$$l_t = \sqrt{\rho_{\max}^2 - \left(\rho_{\max} - 2f_{n_{br}}\right)^2} - f_{az_{br\max}}; \tag{4}$$

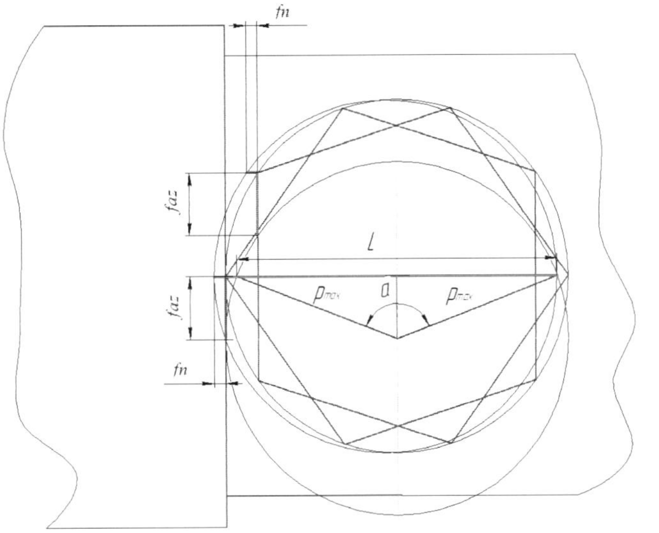

Figure 2. Scheme of destination feed

The azimuthal feed is assigned based on the following relationship:

$$f_{az\max} = \sqrt{\rho_{\max}^2 - \left(\rho_{\max} - f_n\right)^2} = \sqrt{2\rho_{\max} \cdot f_n - f_n^2}; \tag{5}$$

With the known azimuthal feed is necessary to calculate a longitudinal feed, which is defined by the following equations:

$$f_{n\min} = \rho_{\max} - \sqrt{\rho_{\max}^2 - f_{az}^2} = \rho_{\max} - \frac{L}{2}; \tag{6}$$

The chord L, can be determined by the formula:

$$L = 2 \cdot \rho_{\max} \cdot \sin(\alpha / 2); \tag{7}$$

The angle α is expressed as:

$$\cos(\alpha/2) = \frac{f_{az}}{\rho_{\max}}; \tag{8}$$

Based on the formulas, it can be concluded that the rotation speed of the tool V_r should be at an order higher than the summary of the azimuthal feed f_{az} and longitudinal feed f_n.

Apart from the aforementioned main numerical and analytical methods, several experimental studies also investigate size effect of cutting conditions on surface roughness. Three roughness parameters, average roughness (Ra), mean peak-to-valley height (Rz), and maximum roughness (Rmax) by profilometer MarSurf M300 were used to evaluate the roughness. Choice of technological factor levels is carried out by means of single-factor experiment. The results presents on Fig 3.

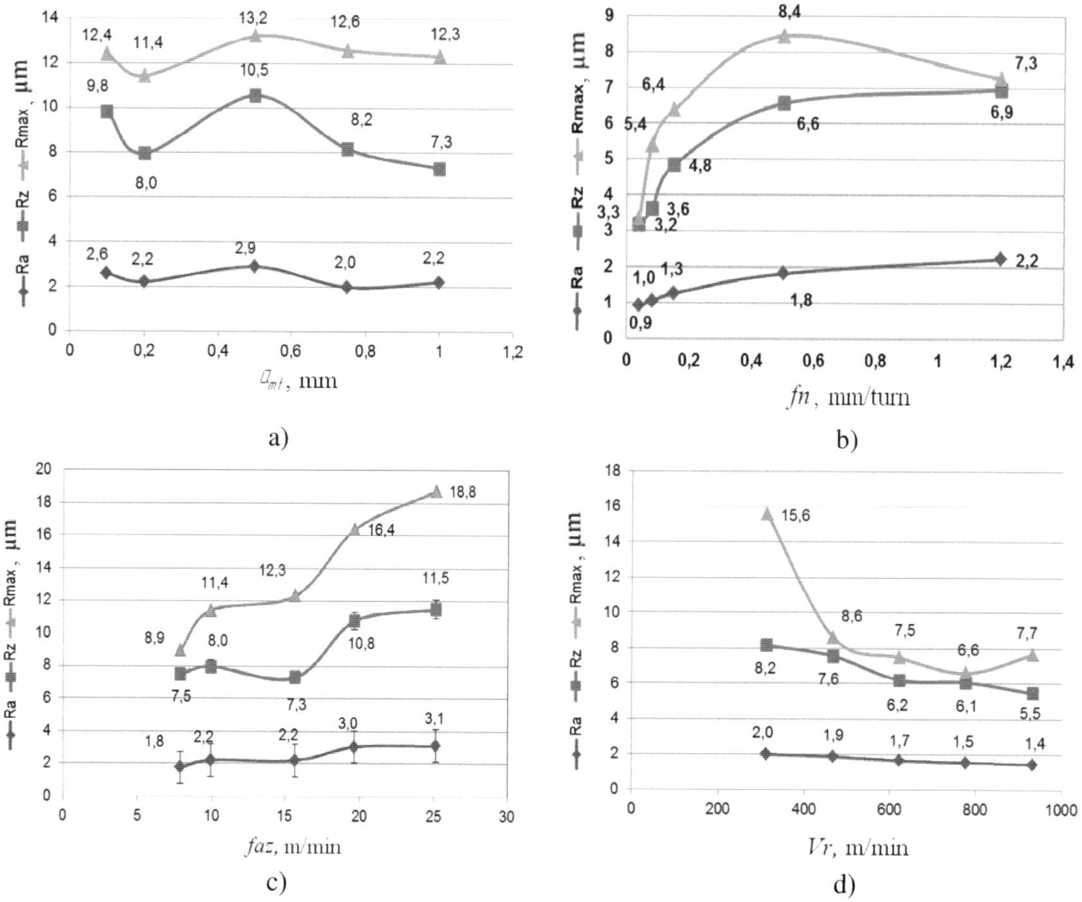

Figure 3. Effect of cutting conditions on surface roughness: a) – depth cutting a_{mf} b) – longitudinal feed f_n c) - azimuthal feed f_{az} d) – cutting speed V_r

Based on the analysis of single-factor experiments were selected varying levels of independent variables for the staging of the planned experiment f_n = 0.3 ... 1.2 mm / turn; f_{az} = 9.891 ... 19.625 m / min; V_r = 310 ... 621 m / min; a_{mf} = 0.3 ... 1.5 mm. Full factorial experiment planning matrix of type 2^4

Processing of experimental data allowed obtaining empirical equations between roughness parameters and cutting conditions (technological modes).

$$Ra = 6.01\frac{f_n^{0.2701}\cdot f_{az}^{0.4679}}{V_r^{0.3081}\cdot a_{mf}^{0.0548}}; \quad (9)$$

$$Rz = 41.52\frac{f_n^{0.2471}\cdot f_{az}^{0.3583}}{V_r^{0.3733}\cdot a_{mf}^{0.0721}}; \quad (10)$$

$$R\max = 251.89\frac{f_n^{0.2208}\cdot f_{az}^{0.594}\cdot a_{mf}^{0.02}}{V_r^{0.6775}}; \quad (11)$$

In order to determine the effect of each of the independent factors on surface roughness was carried out their ranking Fig. 4.

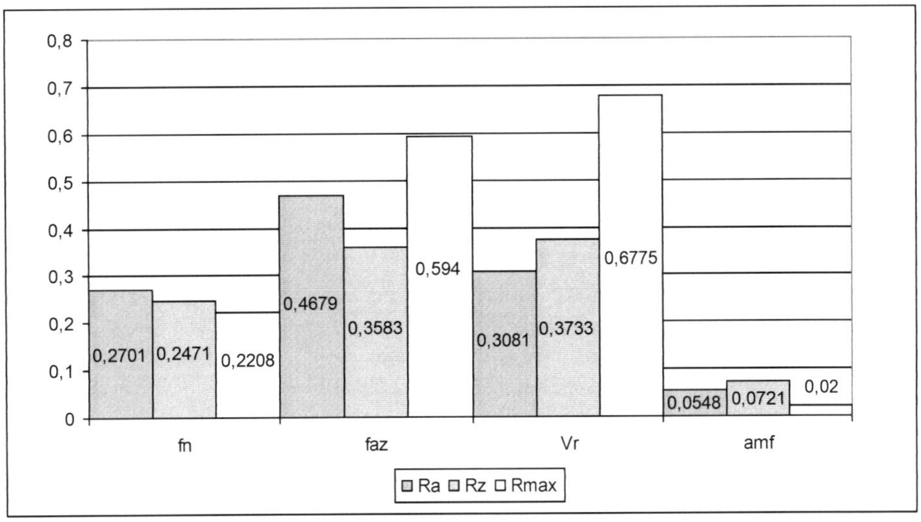

Figure 4. Effect of cutting conditions of RTMC on roughness parameters of machining surface.

Analysis rank chart shows that the dominant influence on the formation of high-altitude surface roughness parameters (Ra , Rz , $R_{\max}$) have azimuthal feed f_{az} and the speed of the main motion V_r. A cutting depth a_{mf} have no effect on surface roughness parameters.

3. Conclusion

Using different cutting edge geometry and the corresponding cutting depth a_{mf} , feeding f_n, azimuth feed f_{az}, cutting speed V_r, it is possible to reduce the contact time of the tool with the workpiece, as well as the frictional forces and the temperature of the tool.

Adhesions are also reduced utilizing the tool inclination angle control and MQL. Thus, our comprehensive theoretical estimates and experimental studies have established the nature of the influence of cutting conditions for RTMC on the roughness of the treated surface. The resulting semi-empirical dependences allow the appointment of the cutting conditions to predict roughness parameters Ra, Rz, $R_{\max}$ (set the cutting conditions with optimal "performance - quality of processing").

4. Acknowledgements

This work was supported by the Russian Found of Fundamental Investigations (grant RFFI № 14-08-00508)

References

[1] Kozhevnikov D V and Kirsanov S V 2006 Rezanie materialov (Cutting Processes) *Mashinostroenie*

[2] Armarego E J A, Karri V, Smith A J R 1994 Fundamental studies of driven and self-propelled rotary tool cutting processes—I,2. Theoretical investigation *International Journal of Machine Tools and Manufacture* vol **34** №. 6 pp 785-801

[3] Yashcheritsyn P I, Borisenko A V, Drivotin I G and Lebedev V Ya 1987 Rotatsionnoe rezanie materialov (Rotary Cutting of Materials) Minsk: Nauka i Tekhnika

[4] Zhang C L et al. 2012 Experimental investigation on surface roughness of KDP crystal processed with rotary ultrasonic face milling *Key Engineering Materials* vol 499 pp 223-228
[5] Yamamoto, Hiromasa, et al. 2009 Thermal Behavior and Chip formation on Rotary Cutting of Difficult-to-cut Materials Utilizing Multi Tasking Lathe and MQL *Proc. 5th Int. Conf. on LEM21* vol **661**
[6] Pan J Y et al. 2014Investigation on surface roughness in orthogonal turn-milling of titanium alloy *J Nanjing Univ Aeronaut Astronout* vol **46** №. 5 pp 720-725
[7] de Oliveira F B et al. 2015Size effect and minimum chip thickness in micromilling *International Journal of Machine Tools and Manufacture* vol 89 pp 39-54
[8] Indakov N S, Binchurov A S 2014 Turning by multifaceted cutters *Russian Engineering Research*, vol **34** №. 1 pp 52-54
[9] Indakov N S, Binchurov A S 2014 Geometry of multifaceted rotary cutters Russian Engineering Research vol **3**4 №. 2 pp 79-82
[10] Binchurov A, Indakov N, Gordeev Y 2016 Method of Rotational Turning With Multifaceted Cutters *IOP Conference Series: Materials Science and Engineering*. vol **124** №. 1 pp 012150
[11] Indakov N S, Gordeev Y I, Binchurov A S, Kiselev D I, Lepeshev A A, Vadimov V N 2016 Influence of kinematic and technological parameters of rotational turning with multifaceted Cutters on deformation processes in the layer workpiece and roughness of treated surface *Vestnik SibGAU*
[12] Gordeev Y I, Abkaryan A K, Binchurov A S, Jasinski V B 2014 Design and Investigation of Hard Metal Composites Modified by Nanoparticles *Advanced Materials Research* vol **1040** pp 13-18.

The device with fluxgate sensor for joint tracking

V Braverman, V Bogdanov, V Belozertsev and O Rozanov
Reshetnev Siberian State University of Science and Technology,
31 Krasnoyarskiy Rabochiy av., Krasnoyarsk, 660037, Russia

E-mail: braverman-vladimir@rambler.ru

Abstract. The possibility to evaluate electron beam deflection from joint by magnetic field strength of the current flowing through the welded part of joint is studied. The differential flux gate is utilized as a sensor of electron beam and joint displacement. The concept of the proposed way of electron beam position control relatively to joint is that welding current is divided into two components. Value of beam deflection is defined due to value and direction of vertical magnetic field strength component. At the same time magnetic field strength over welded part of joint is measured.

Introduction. Since the moment when electron beam welding was applied in industry first time the large experience of such type of welding has acquired. On the other hand the problems connected with assurance of accuracy of beam position along joint have not had perfect solutions. This task is a subject of the paper and other issues [1–7]. One reason of poor quality of welded joints is low accuracy of automatic tracking system that directs electron beam along joint. There are many mentioned control systems where sensor is located with some distance from welding place. Therefore deflection between sensor is measured and result has difference from deflection between sensor and joint where last one is needed. Due to such measuring conditions the inaccuracy of overlapping joint and beam can appear in case of curvilinear joint, for example. Moreover, there is high level of electromagnetic interference, which disturbs electronic control system operating. The analysis has shown the following. To improve accuracy of the beam control system information about beam and joint positional relationship can be extracted from electromagnetic field behavior, as electromagnetic processes accompany welding process.

The way of beam and joint positional relationship definition. Proposed technique is based on magnetic field of welding current (I_B) and beam coordinates identification. Principle of this way is that under beam deflection from the joint redistribution of welding current components and current-induced magnetic fields happens [8, 9].

Due to current collectors electron beam current I_B is divided into two components I_1 and I_2 (Fig. 1). In case when electron beam is located strictly over the joint then magnetic-fields strength H_1 and H_2 have equal values and opposite directions (Fig. 1, *a*). In such case total magnetic field is defined by current I_B, and vector of magnetic-field strength is located in horizontal plane. Under beam deflection (-ε) from the joint (Fig 1, *b*) current I_2 flows through welded part of the joint, and magnetic-field strength produced by current I_2 is changed on value H_V. The total magnetic field over welded part of the joint is defined by vertical component of magnetic field strength H_V from current I_2, flowing through welded part of the joint, and horizontal component of magnetic field strength from beam current I_B. In case of beam deflection to opposite direction, current I_1 flows through the welded part of the joint and magnetic field strength produced by current I_1, is changed on value $-H_V$ (Fig. 1, *c*). So the total magnetic field over welded part of the joint is defined by vertical component of magnetic field strength $-H_V$ from current I_1, flowing through welded part of the joint, and horizontal component of magnetic field strength from beam current I_B. Therefore vertical component of magnetic field strength over welded part of the joint gives information about beam position relatively to joint. In case of magnetic field sensor (for example flux-gate meter) utilizing under condition that sensitive axis located in plane of the joint but perpendicularly to plane of weldment, vertical components H_V and H_V can be measured. It is sufficient information to define electron beam position.

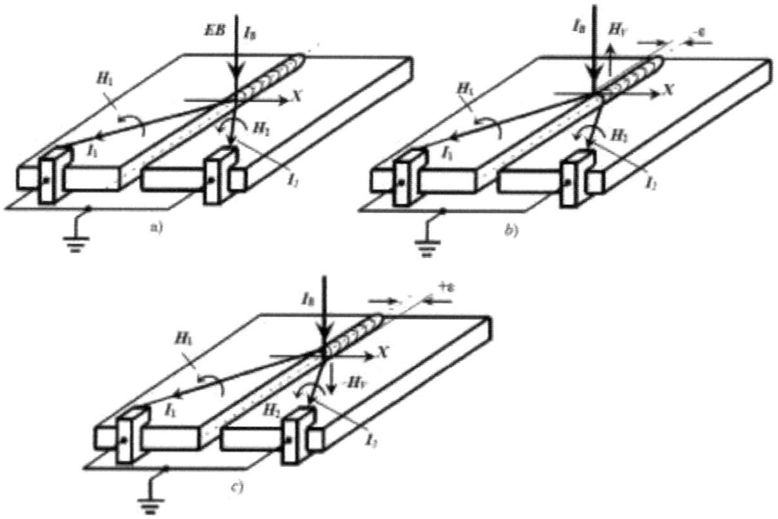

Figure 1. Dependence of magnetic field vertical component on electron beam position relative:
a) – X=ε=0 (no displacement); b - X=-ε (left displacement); c – X=+ε (right displacement)

Experimental analysis. The magnetic field alteration depending on electron beam position relatively to the joint is proved with help of physical model. In the capacity of conducting medium (the simulation of weldments material) electrolytic solution is used. The joint was simulated by dielectric sheet. So voltage source, current contacts, electrolytic solution and electrode drowning to electrolytic solution in border between joint and weld generate model of welding current electrical network.

With the help of the probe and meter the family of equipotential lines on the surface of electrolytic solution has been received.

Thereto it is necessary to define set of points with equal potential as Fig.2 shows. Then current lines of force are plotted taking to account that lines of force must be normal to equipotential lines in each point (Fig. 2).

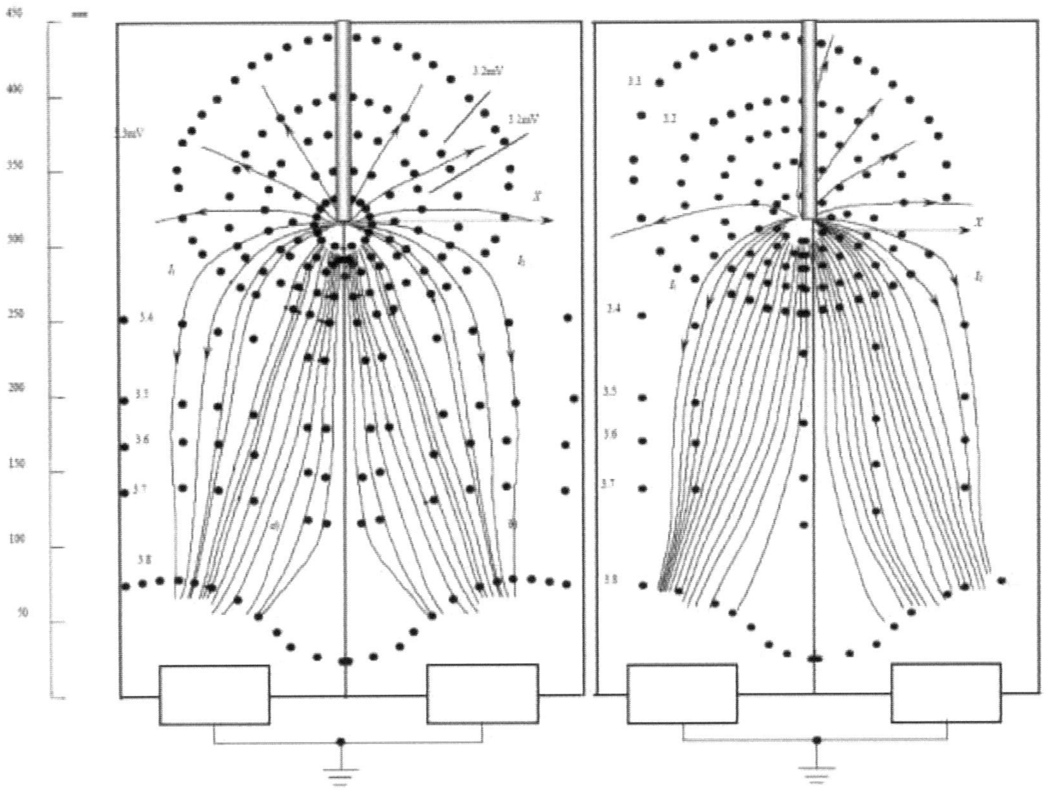

Figure 2. Physical simulation results of current distribution I_1 and I_2: a) - ε = 0; b) - ε = 0,1 mm

As Fig 2,b shows, under beam deflection from the joint the symmetry of current distribution is disturbed. Therefore, total magnetic field differs from the field under coincidence of beam and joint positions. (Fig. 2, *a*).

Theoretical analysis. Analytical definition of magnetic field strength dependence from electron beam deflection relatively to joint was done with 3*D* model, which allows simulating of welding process from the position of electromagnetic properties of space restricted in plane of weldment by coordinates that were defined due to two-dimension model analysis.

Mentioned dependence can be presented as:

$$H_V = F(x, y, h_0, I),$$

where H_V – vertical component of magnetic field; *x*, *y* – beam coordinates; h_S –height of the flux-gate meter; *I* – welding current.

The simulation is based on Maxwell equations calculation:

$$rot\vec{H} = \vec{j} + \frac{\partial D}{\partial t},$$

$$rot\vec{E} = -\frac{\partial \vec{B}}{\partial t},$$

$$div\vec{D} = \rho,$$

$$div\vec{B} = 0,$$

where E – electric field strength, D - displacement field (electric flux), H – magnetic field strength, j- current density, ρ – volume charge density, B – flux density.

Maxwell equations are added with constitutive equations that connect vectors E, D, H and B to quantities characterizing electrical and magnetic medium properties:

$$\vec{D} = \varepsilon_0 \varepsilon \vec{E},$$

$$\vec{B} = \mu_0 \mu \vec{H},$$

$$\vec{j} = \gamma \vec{E},$$

where ε – relative dielectric constant, μ – relative permeability, γ – electric conductivity, ε_0- electric constant, μ_0 – space permeability. The current injects to the network through the beam and is taken off from the side of non welded part of the joint.

The weldment is zoned by cylindrical air layer. In this region magnetic field should be calculated under x coordinate of the beam variations. Obtained calculations were produced under following assumptions: the medium is isotropic, ε and μ parameters are real numbers. The voltage is applied to upper edge of the element - "beam" to provide welding current 250mA.

Two parameters were modeled - direct current in conducting medium and electromagnetic field. The equations were calculated due to finite elements method.

The calculations were produced for aluminum alloy and steel materials under beam deflection variations. In Fig. 3 results of calculations for welding process of parts from aluminum alloy with thickness 10 *mm* under I_B = 250 *mA* and speed of welding process S_W = 0,5 *sm*/*s* are presented. The lines show currents and full tones correspond to distribution of magnetic fields.

It shows that in case when beam deflection is zero, the distribution of welding current components I_1 and I_2 and magnetic fields produced by them are symmetrical relatively to the joint (Fig. 3, *a*).

In case of beam deflection from the joint the symmetry is disturbed (Fig 3, *b*), and some current component flows through the welded part. The stronger the symmetry is disturbed, the larger the beam deflection is. The same effect is occurred under the physical simulation that was described earlier. As a result the balance of currents I_1 and I_2 is disturbed. The vertical component of the magnetic field H_V appears. The value H_V depends on value and direction of beam deflection from the joint (Fig 3).

All in all, the calculation results have shown that beam deflection from the joint converts to deviation of vertical component of magnetic field.

The dependence H_V from ε can be represented approximately as

$$H_V \approx \frac{1}{4\pi r_0} \cdot \frac{I_B \cdot \varepsilon}{\sqrt{r_0^2 + \varepsilon^2}},$$

where r_0 – average distance from magnetic field source to point of measuring, I – average beam current value. The calculations have shown, that magnetic field strength H_V is almost proportional to beam deflection from the joint (Fig. 4).

The research results authenticate the possibility of beam position definition relatively to joint by value and direction of magnetic field vertical component. Meanwhile, the sensor (flux gate) can be located with some distance from the beam but it doesn't lead to procedure error of measurement, as vertical component of magnetic field H_V appears only under beam deflection from the joint.

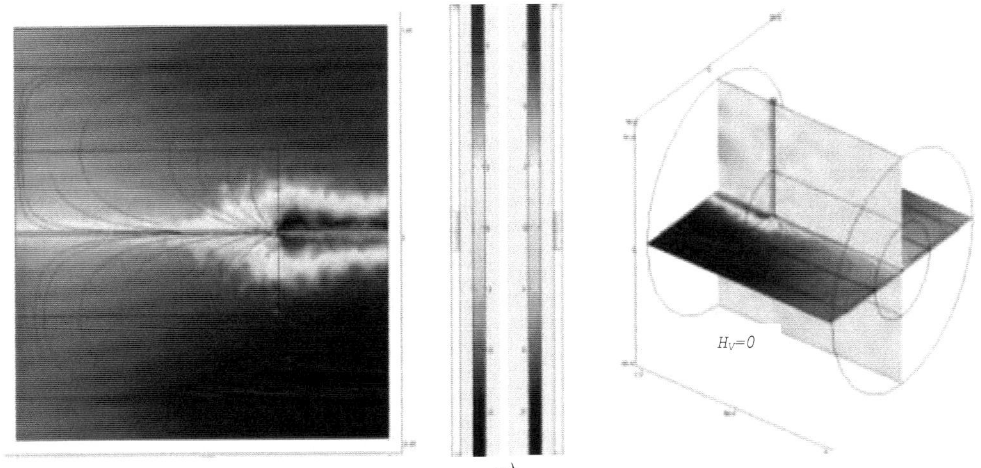

a)

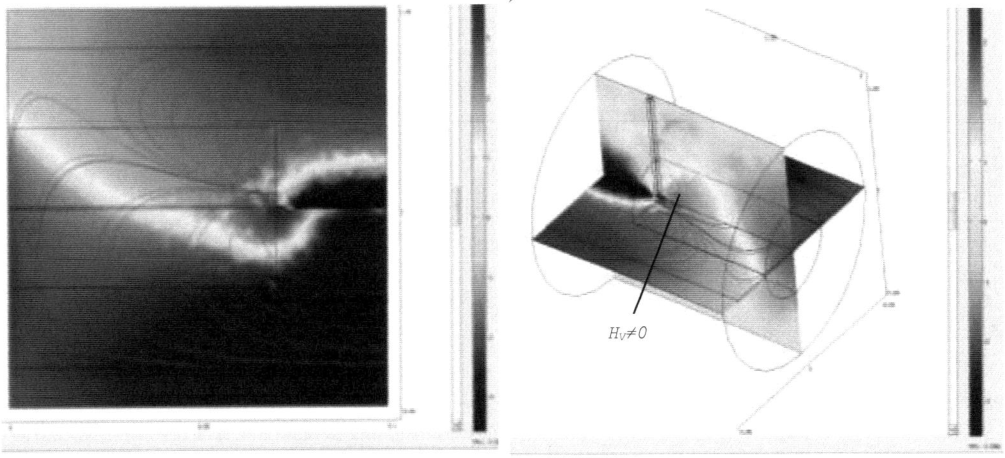

b)

Figure 3. Magnetic fields distribution: a – ε =0; b – ε =0,1mm

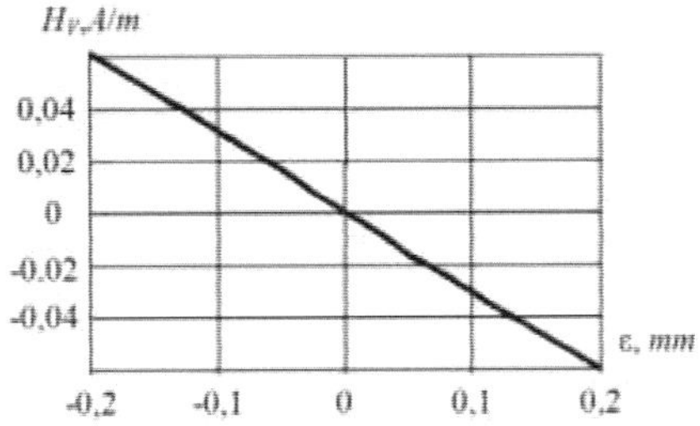

Figure 4. Dependence of magnetic field vertical component from beam deflection relatively to joint

The system for joint tracking. In Fig. 5 the simplified diagram of the system for automatic joint tracking with flux gate as a sensor of beam position relatively to joint of welding parts is given. The system consists of flux gate *S*, converter-amplifier *T* to provide standard electrical signal from the flux gate, beam-deflection system *DS* to shift beam position in the line perpendicularly to joint line.

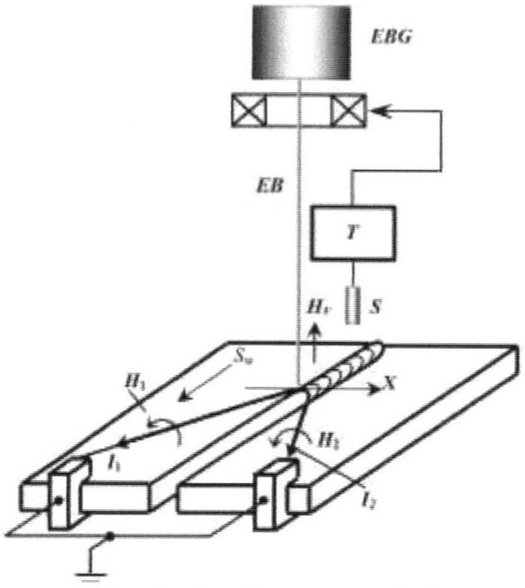

Figure 5. Structural diagram of the device for automatic beam control: *S* – sensor; DS_X – deflection system for *X*-axis;
T – transducer; *EB* – electron beam; *EBG* - electron beam gun

The sensor is fixedly connected with electron-beam gun under welded part of the joint. Sensitive axis is located in plane of the joint but perpendicularly to plane of weldment. When the beam is located on the joint (Fig. 1, *a*), sensitive axis projection of the magnetic field strength is produced by welding current under welded part of joint equal to zero. So output signal from the sensor is also equal to zero.

Under beam deflection from the joint (Fig. 1, *b*) component of current welding I_2 flows through the welded part of joint. So, sensitive axis projection of the magnetic field strength characterizes value and direction of beam deflection from the joint. In this case the measuring coil of the flux gate generates the signal that is converted and gained by device *T* and becomes *DC* voltage that is proportional to beam deflection value from the joint. The mark of voltage defines direction of the beam deflection. The voltage signal enters the input of deflection system *DS*, which compensates displacement of beam location from the joint.

The proposed system was tested in real conditions and received error of beam and joint overlap was less than 0,2mm.

Conclusions

1. The method of electron beam location relatively to joint, based on identification of magnetic fields and coordinates of the electron beam, is proposed.
2. The proposed technique allows controlling of electron beam location and correcting it with help of automatic system for joint tracking with low error.
3. The noise immunity of the proposed system is very high, because in the proposed system the welding current is information source whereas in the known similar systems the welding current is the source of electromagnetic noise.

IOP Conf. Series: Materials Science and Engineering **255** (2017) 012002 doi:10.1088/1757-899X/255/1/012002

References

[1] Reichmann A, Leffler D, Bartel R 2004 State of the art of a FEP control unit of beam scanning for electron beam equipment *6th International Conference on Beam Technology*, Halle (Saale)

[2] Braverman V, Shabanov V, Bashenko V and Bajakin S 2004 Control over electron beam welding process by X-ray radiation from the zone of welding *6th International Conference on Beam Technology*, Halle (Saale)

[3] Braverman V 2008 Decelerating X-ray radiation during electron beam welding and its interrelation with the process parameters *Vestnik SibGAU* vol **3**

[4] Braverman V, Belozrtsev V, Litvinov V and Rozanov O 2008 Issues relating to controlling seam formation process in electron beam welding *Vestnik SibGAU* vol **2**

[5] Braverman V, Belozrtsev V, Goryashin N and Lelekov A 2009 Mathematical models for the process of changing the x-ray radiation from the beam position relatively the seam during the electron beam welding *Vestnik SibGAU* vol **2**

[6] Braverman V, Bashenko V and Bajakin S 1997 Device for control focus and depth of penetration on his own X-rays during electron beam welding in modulation of focus Welding Production vol **1**

[7] Braverman V, Shabanov V, Bashenko V and Bajakin S 1998 Method of electron-beam welding. Patent of Russia N2113954, International Classification Code B23K 15/00

[8] Braverman V, Laptyonok V, Belozertsev V 1982 Fluxgate sensor for joint tracking. Patent of USSA N941056 USSR, international classification Code B23K9/10

[9] Braverman V, Laptyonok V, Belozertsev V 1982 Method for measuring deviations of the electrode from the seam. Patent of USSA N941057 USSR, international classification Code B23K9/10

IOP Conf. Series: Materials Science and Engineering **255** (2017) 012003 doi:10.1088/1757-899X/255/1/012003

Mechanical fasteners used in historical Siberian shipbuilding: perspectives for metallurgical analysis*

A E Goncharov[1], D M Mednikov[2], N M Karelin[3], I R Nasyrov[4]

[1] Associate professor, Reshetnev Siberian State University of Science and Technology, Krasnoyarsk, Russia

[2] Lecturer, Reshetnev Siberian State University of Science and Technology, Krasnoyarsk, Russia

[3] Senior lecturer, Reshetnev Siberian State University of Science and Technology, Krasnoyarsk, Russia

[4] Post-graduate student, Reshetnev Siberian State University of Science and Technology, Krasnoyarsk, Russia

E-mail: goncharovae@sibsau.ru

Abstract. Recent discoveries of shipwrecked vessels in the northern reaches of the river Yenisei led to a number of questions concerning the history of shipbuilding in Siberia and the technical features of the first vessels of the industrial era to navigate the Northern Sea Route and the Yenisei. One of these questions addresses the features of mechanical fasteners used in the construction of the Siberian vessels. The answer to this question may provide information on how the first vessels, constructed in Siberia during the 1870's, were able to sail the high seas of the Arctic Ocean and reach European ports. In this paper, we provide a description of iron mechanical fasteners obtained from one shipwrecked vessel and discuss on the perspectives of a metallurgical analysis

1. Introduction

Recent expeditions to the Siberian Arctic, both scientific and amateur, have revealed a number of unidentified shipwrecked vessels [1]. Unfortunately, the problem of historical shipbuilding in Siberia has yet been properly addressed in research literature. Samples of mechanical fasteners from the discovered ships, the majority of which are barges, may shed light upon this issue.

This paper addresses the problem of historical shipbuilding in Siberia in the latter half of the nineteenth century and the early half of the twentieth century, specifically focusing on the mechanical fasteners used for the construction of these vessels. A short history of shipbuilding in Siberia during Russia's era of industrial development is given, the field research of the historical expedition, which collected and examined many of the samples of mechanical fasteners is reviewed upon. The paper then proceeds to discussing the prospects of conducting a metallurgical analysis of the collected samples, describing the procedure of such an inquiry and stating the predictions.

*The reported study was funded by Russian Foundation for Basic Research, Government of Krasnoyarsk Territory, Krasnoyarsk Region Science and Technology Support Fund to the research project № 16-11-24010.

Due to space limitations, the authors shall focus, primarily on shipbuilding on the Yenisei. This river was more important to navigation on the western leg of the Northern Sea Route as it could be easily accessed by large oceangoing vessels, unlike the neighboring Ob'. This fact stimulated the development of marine shipbuilding on the Yenisei. However, naval architecture was similar throughout all of Siberia, as it will be shown in this paper, making shipbuilding on the Yenisei a potent example for further research.

2. Shipbuilding in Siberia. Results of the 2016 expedition

Historical literature offers very little information on the subject of Siberian shipbuilding. The topic is briefly considered in a number of studies on the history of the Northern Sea Route and the development of navigation on the great Siberian rivers, the Ob', the Yenisei, and the Lena, however, to our knowledge, no inquiry has yet been made into the technical features of the vessels constructed locally. Even the more prominent vessels, which made their way into the annals of the history of the Northern Sea Route, the clipper-boat *Sevenoe siianie* (Aurora Borealis) and the schooner *Utreniaia zaria* (Morning Dawn), continue to remain unknown in terms of their technical features. An essay by E.N. Vladimirov, published in 1940, continues to remain the only potent publication on the technical characteristics of these two vessels. Vladimirov specifies that the first vessel was 90 feet long, 14 feet wide, and 8 feet high [2]. No further information is provided and the author fails to cite any source. The second vessel is described on the basis of a note, written by the shipwright of both vessels, the Heligolander Petr [Peter] Boiling. The schooner is described as having a length of 56 feet, a width of 14 feet, and a height of 7 feet. The displacement is 50 tons. The vessel was built from a variety of woods: larch was used for the bottom and hull draft; pine and fir were used for the rest of the ship [2]. The skipper of both vessels provides some additional information on the features of the *Severnoe siianie*, specifically its tonnage – approximately 100 tons, and some details on the rigging [3].

The vessel was lost with most of her crew during the winter and spring on 1876–1877. In 2016 an expedition was launched by Reshetnev Siberian State Aerospace University (SibSAU) and the Russian Geographical Society to find the wreck of the 140-year old ship in the delta of the Yenisei. A wreck was found on the island Bol'shoi Brekhovskii in the Brekhovskie ostrova archipelago (wreck A). As it was determined that it was on this island that the crew of the lost vessel had wintered, it was proposed, on site, that the wreckage could be the clipper-boat *Severnoe siianie*. The dimensions of the wreck matched the description, given in Vladimirov's essay. After a press conference with the expedition team, Russian and international mass media reported the find to be the nineteenth-century vessel.

Further investigation showed that the vessel was actually much younger than the proposed ship. This immediately posed a problem as there is very little evidence on both the subject of shipbuilding in Siberia and wrecked vessels in the Yenisei delta. The expedition found another vessel at the settlement of Kalaul (wreck B). From its size and preserved features, it was determined that this was a wrecked barge or lighter. Some features of this vessel's design, particularly the mechanical fasteners used, were very similar to the ship found of the Bol'shoi Brekhovskii ostrov.

It has been suggested that both vessels were built between the 1930's and 1950's. From archival materials we know that the majority of barges and lighters for needs of the Northern Sea Route have been built at Krasnoyarsk and Pridivinsk [3; L. 153]. By the mid-1930's it was becoming increasingly evident that it was necessary to begin vessel production at Igarka [5, L. 112]. Another facility on the Yenisei that had its fleet of barges was the port of Dudinka. By 1939, Dudinka had 28 barges with a gross tonnage of 21,200 tons (750 tons of tonnage for each barge) [6, L. 106].

It is known from both archival sources and historical literature that rudimentary shipbuilding existed on the Yenisei beginning from the seventeenth century. However, vessels that could operate during continuous periods in the mouths of the great Siberian rivers and in the seas of the Arctic Ocean were built on a large scale beginning only from the late 1800's. This is especially true for the Yenisei, where river navigation was less intense than on the neighboring Ob'. At the same time, the experience of both traditional Russian shipbuilding and the construction of European-style ships for the Great Northern Expedition (1733–1743), which was an exceptional event, had little impact on the further development of local river shipbuilding. Considering that navigation in the Kara Sea to the mouths of the Ob' and Yenisei was first performed only in the 1870's, it is obvious that there was no naval engineering until this date. The slow but steady progress in the development of the western leg of the Northern Sea Route during the late 1800's and early 1900's advanced local shipbuilding, particularly the construction of barges and lighters. However, the remoteness of the first Siberian wharves made them extremely difficult to supply with industrial materials, primarily, sheet metal, iron structural elements, and mechanical fasteners. This made wood the material of choice for shipbuilding in Siberia until the 1950's. However the abundance of wood could not compensate the need for metal fasteners, which either had to be delivered as completely manufactured items, or as semi-finished products or raw material that would be processed locally.

Figure 1. The structure of a barge hold (cargo box) as seen in a grounded barge wreck in the Laptev Sea (c. 1930's–1950's). Image from the 2014 expedition of Sergei Karpukhin (accessed http://karpukhins.livejournal.com/181715.html) (printed with permission of author)

Another problem is the dating of objects and structures that are between 150 and 80 years old. Typically, this problem does not occur as there is enough documental evidence reporting on the date a concrete object was built or lost. However there are still many cases when it is extremely difficult for the researcher to find the necessary historical records. Over the last century and a half many documents have been lost or disposed of as not 'having historical value'. Not being able to find reference in archives to the construction or destruction of a structure or object, researches turn to the local testimony. Very often, however, the locals are simply not aware that there is a historical site in the area, do not know its history, or provide inaccurate or exaggerated information on the object or site. Thus, such information can only be used if it is cross-examined with more reliable materials.

An example to illustrate the aforementioned point has been recorded during the 2016 expedition. The team was working on the river Sal'naia kur'ia, a tributary of the Yenisei, searching for the British steam-schooner *Thames*, which sank there in 1878. The locals confirmed that there was a wreckage of a vessel on the bottom of this river; however they were surprised to learn that this was a nineteenth century vessel, supposing that is was an ordinary motor boat (Russ. – *kater*) [7].

3. Types of mechanical fasteners used in Siberia shipbuilding

The 2016 expedition documented and photographed mechanical fasteners from both vessels found. Three samples of wrought iron nails (spikes) were retrieved from wreck A to conduct a laboratory investigation of its properties. In this section we shall review on the types of mechanical fasteners used in the construction of these two vessels and on the properties of wrought iron.

An intense study of the properties of iron fasteners (rivets) is given in [8]. Hooper et al. gives a definition of wrought iron as a commercially pure iron (less than 0.02 wt% C) with slag content of 1–2 wt% [8]. The puddler process was the chief production method of wrought iron during the late nineteenth and early twentieth century. Wrought iron would be the most available material for the construction of mechanical fasteners in Siberia until the industrialization of the 1930's.

Table 1. Assortment of mechanical fasteners found on the two vessels

Type	Dimensions	Description
Nails (spikes)	Type A: l = 25 cm, w = 0.8 cm (at head)	Found in great number on wreck A. Used for attaching deck planks to the transverse beams of the frame. All collected samples are deformed (bent).
	Type B: l= 35 cm, w = 1 cm (at head)	Found in great numbers on both vessels. Used for attaching hull planks to the frame of the vessel. Many fasteners remain intact, holding outer hull planks to frame elements.
Bolts	l = 9 – 40 cm, ⌀ 1.2 cm	Most bolts are square-headed, singular hexagonal-headed bots can be found. Bolts were the essential fastener for building the vessel's frame. The deformation of the bolts is clearly visible. Found on both vessels.

Nuts	w = 3–4 cm	The prevailing type of nut is square. Singular hexagonal nuts can also be found. Little signs of strain can be seen. Found on both vessels.
Washers	w = 2–4 cm	All washers are either square or rectangular. Found on both vessels.
Threaded rods	l = 2–4 m, ⌀ 1.2 cm	Their function is not clear. Found only on wreck A.
Turnbuckles	l = 40–60 cm	Turnbuckles were used to connect two threaded rods and to create tension. As with the threaded rods, their function is not clear. Found only on wreck A.

An investigation into the types of mechanicals fasteners provided clues on the construction date of the two vessels. The initial supposition that wreck A was the clipper-boat *Sevenoe siianie* proved to be erroneous. Besides the dimensions of the discovered wreck matching the ones of the nineteenth century vessel, it was considered that hand-wrought-iron nails (spikes) could not be used in the construction of vessels, produced after the 1800's. An inquiry into the history of nail production and their usage emphasizes on using nails to date old buildings; with the introduction of machine-cut nails in the late eighteenth and early nineteenth centuries, and wire nails, which became the dominant nail type beginning from the 1890's [9]. The nails, recovered from the first vessels (3 specimens), are of Type A (2 specimens (Figs. 2–3)) and of Type B (1 specimen (Fig. 3)). They are hand-wrought, being square in cross-section. The heads are produced by a heading tool. The head diameters vary from 4 to 4.5 cm.

As seen from the images of the wreck, nails were the most distributed mechanical fastener used to construct these vessels. Their implementation ranged from securing the bottom planks of the below waterline hull to the floor beams, securing the hull planks to the ribs, and fastening the deck planks to the deck beams. The use of such crude nails in Siberian shipbuilding lasted deep into the twentieth century. Apparently, industrial production of such large nails lagged behind the growing demand for them as the shipping on the Northern Sea Route became ever more intense. Local nail production, however obsolete, provided a temporary solution to the problem. However, in the 1930's it was also incapable of coping with the demands for the construction of new vessels and the maintenance of existing ones as seen from archival materials [5, L. 60 verso].

Figure 2. Nails from overturned deck planks. Signs of deformation caused by bending and torque are visible

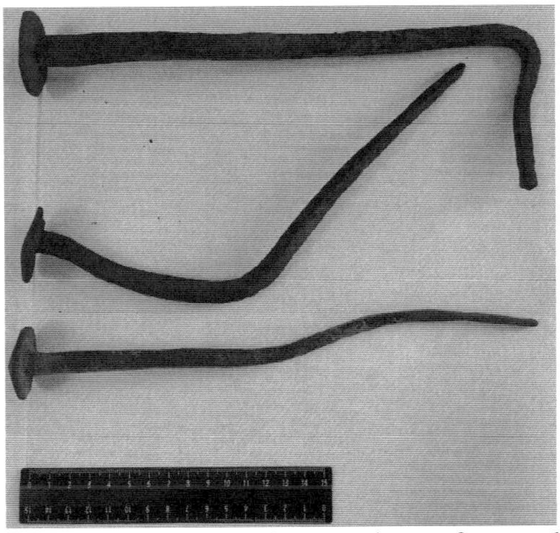

Figure 3. Three nails taken from the first wreck as specimens for metallurgical analysis. Top – Type B; bottom (2) – Type A. All specimens were bent. The first nail was bent deliberately for additional structural strength; the second specimen demonstrates deformation caused by torque (apparently the wooden member was twisted around the axis of the nail)

Bolts, as seen from Table 1, played a critical role in the assembly of the ship frame (Figs. 4–5). The vessels, both of which were large river barges were flat-bottomed, which, according to structural mechanics, requires a stronger frame than curved or angular bottoms. The members of the frames of both vessels consist of ribs (15 cm thick), deck beams (not preserved), cross-members (diagonal beams), which provided extra support for the ribs, and floor beams. All of these structural elements were held together by bolts. The washers that were fitted on the bolts and provided load distribution over the surface of the wooden members were all crudely made from lengths of iron or steel strips and cut to size. The majority of nuts were also crudely made, being square and even rectangular, although separate hexagonal nuts can still be seen on the wreck; these, most likely, were imported from the industrial centers of European Russia.

Figure 4. Two bolts, square washer, and square nut with scale marker. These two samples were bent

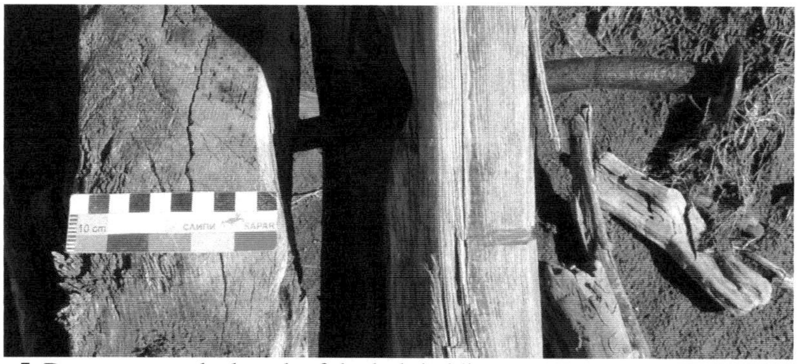

Figure 5. Demonstrates the length of the bolt holding two wooden members of the ship frame. The bolt and the wooden members are deformed due to shear stress

The most curious mechanical fasteners found were threaded rods and turnbuckles on wreck A (Figs. 6–7). No analogs have been found on other wrecks. It was first supposed that the rods were either alien to the structure of the vessel (installed years later to transport the wreck, supposing that it was an obstacle to navigation), or were somehow utilized to secure cargo. However, it seems more likely that these were initial structural elements of the vessel. The rods and turnbuckles supposedly held the mooring posts; they were pinned to the floor of the barge and stretched to the deck, thus, giving us information on the vessel's depth. A key feature of this component, allowing us to more or less accurately date the vessel is the design of the turnbuckles. An analysis conducted by the Department of Welding of Flying Apparatuses concluded that the turnbuckles were made by welding two iron plates to two hollow cylinders with an inside thread. The type of welding used was arc welding. Fig. 6 is a testament to this observation; traces of the weld pool are still visible on the structure. Consequently, vessel A could not have been built prior to the 1930's; even at that time there would hardly be enough welding apparatuses in Siberia. It is known, for instance, that among the first recorded incidents of arc welding in the Soviet Union in shipbuilding took place in 1930. Archival evidence let us establish that by 1936 there was at least one electric arc welding apparatus at Igarka; its power consumption was 30,144 kWh [5, L. 36]. So, it is quite likely that wreck A was either built in the 1930's, or some of its components were installed at that time.

Figure 6. A turnbuckle with two threaded rods from wreck A. The weld is clearly seen. The upper plate buckled

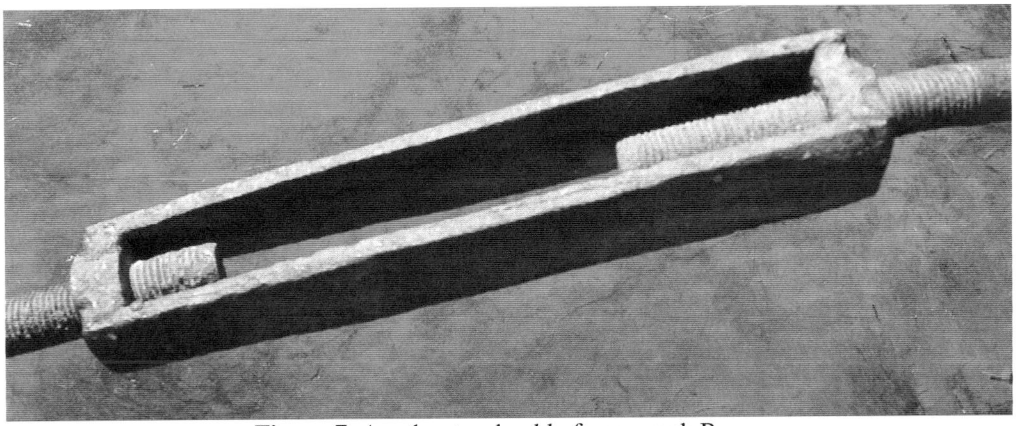

Figure 7. Another turnbuckle from wreck B

4. An inquiry of the types of failures

However effective these fasteners were, they still could not prevent various types of failure resulting in the heavy stress endured by the vessel from excessive loads, harsh sea conditions, etc.

The current condition of the wrecks shows that most of the fasteners are still intact despite being massively deformed. This is, primarily, a result of the wood decay. Some of the surviving wooden members also reveal various cases of failure due to stress. Among the discussed types of failure in this paper will be failure modes for bolted joints and

As we have found no signs of sheared bolts or nails, shear failure of the metal fasteners seems to have been a minor concern. However, it affected the wooden members to some degree; as the vessel rocked on the waves (usually, an up and down motion), shear stress caused deformation around the bolt holes (Fig. 5). Bearing failure of the wooden structures also seems to have been an insignificant concern, due to the properties of wood to withstand

IOP Conf. Series: Materials Science and Engineering **255** (2017) 012003 doi:10.1088/1757-899X/255/1/012003

such deformation; many of the washers are embedded into the wood – a result of overtightening and the expansion of moist wood.

Considering the threaded rods, their deformation can be clearly visible on the first wreck. As the rods are all deformed in one pattern, it can be suggested that this was a result of the deck collapsing and compressing the rods and turnbuckles, which acted as slender columns and buckled. Using Leonhard Euler's critical load equation

$$P_{cr} = \pi^2 \frac{E\,I}{L^2}$$

where I is the smallest moment of inertia for the cross-sectional area of the column, we can find the critical load that causes an ideal column (secured at both ends) to buckle [10]. In our case, the column is pinned at both ends, as the turnbuckles allow it to rotate around its axis, thus, K (the effective length factor) = 1.

Likewise, we can find the load that caused the deformation of the threaded rods, by rewriting Euler's equation

$$P_{cr} = \pi^2\, E\, I\, /\, L^2 = \pi^2\, E\, /\, L^2\, \pi\, d^4\, /\, 64 \rightarrow d = \sqrt[4]{(64\, P\, L^2\, /\, \pi^3\, E)}.$$

Further modifications of Euler's critical load equation can incorporate the effective length factor (K = 1, 0.5, 0.7, 2)

$$P_{cr} = \pi^2\, E\, I\, /\, (KL)^2;$$

the Euler stress can be calculated

$$\sigma_{cr} = \pi^2\, E\, /\, (KL/\, r_G)^2 \text{ [10].}$$

The turnbuckles would be the element that tensile stress would be most exerted on. The welded joints would be the most vulnerable part of the component. However, as seen from the images, either the tension was little, or the joints proved strong enough.

While there is not necessarily anything new in the behavior of these components in terms of structural mechanics, it is possible, by using materials science, to extract some new historical evidence concerning the vessels' construction, its loss, and the environment in which the wreck has been found. Further inquiry into the issue shall be made to solve a number of topical issues.

5. Perspectives for metallurgical analysis

A study of metal fasteners from the RMS *Titanic* and comparing them with other iron components of the period has been discussed in [8]. The paper provides us with a scientific apparatus for performing experiments on the metal fasteners, collected from wreck A. It is planned that the perspective work shall include the following: microstructural examination and image analysis for slag contents (optical microscope), mechanical testing (materials testing machine), and micromechanical analysis. The collected specimens will be compared with other components of the period, e.g. railway spikes from the Trans-Siberian Railway.

It has been ambitiously proposed that a complex metallurgical analysis can even be used to date metal objects. The method can be basically outlined in the following thesis: let us suppose that we have at our disposal two metal components that have been produced at the same time from the same material, and using the same technological process; in this case a high probability exists that these two components are be able to endure similar loads and withstand similar stresses and strains; their microstructures will also be similar. The method is absolutely theoretical and has, to our knowledge, not been proposed in literature. At the same time this dating method, which is based on material properties and strength of materials, offers an alternative to more traditional approaches, such as radio-carbon dating and dendrochronology.

6. Conclusion

The conclusion to be derived is that Siberian shipbuilding during the first decades of the Soviet era was in sufficient deficit of industrial building materials. Mechanical fasteners had to be produced on site using whatever rudimentary methods available or delivered at high costs to the local wharves. Using various methods from structural mechanics and physics, it is possible to obtain new evidence on the structures of the first vessels that were employed in the Kara Sea Expeditions.

Further research is planned after sufficient financial support will be obtained. It is also planned to collect more specimens from shipwrecked vessels on the Yenisei to conduct a complex analysis of the structural members and mechanical fasteners of nineteenth and twentieth-century vessels. The results of these investigations may later be used to reconstruct one of the more prominent vessels – the schooner *Utreniaia zaria*.

References

[1] Goncharov A E, Karelin N M, Mednikov D M, Nasyrov I R 2016 GIS and satellite remote sensing for archeology: exploring polar history *Vestnik SibSAU* vol **4** **(17)** pp 956–963

[2] Vladimirov E N 1940 Geroicheskii reis shkhuny "Utreniaia zaria" [The heroic voyage of the schooner *Utreniaia zaria* (in Russ.)] (Moscow–Leningrad)

[3] Shvanenberg D I 1877 O plavanie iakhty "Utreniaia Zaria" iz Eniseia cherez Karskoe more i Severnii ocean do Varde [The voyage of the yacht Utreniaia Zaria from the Yenisei across the Kara Sea and the Northern Ocean to Vardö (in Russ.)], *Trudy Sankt-Peterburgskogo otdela Imperatorskogo obshestva dlia sodeistviia russkomu torgovomu morekhodstvu (s 8 fevralia 1877 po 1 ianvaria 1878 g.).* (St. Petersburg) pp 244–254

[4] Igarskii port GUSMP in Ekonomicheskaia kharteristika Igarskogo administrativnogo raiona [The port of Igarka of the Chief Administration of the Northern Sea Route in The economic characteristics of the Igarka administrative raion] (1938), Munitsipal'nyi arkhiv g. Igarki [Municipal archive of Igarka] (F. P-2. Op. 1. D. 255)

[5] Ekonomiko-geograficheskii obzor Igarskogo raiona [An economic and geographical review of the Igarka raion (in Russ.)] (1935), Munitsipal'nyi arkhiv g. Igarki [Municipal archive of Igarka] (F. P-2. Op. 1. D. 254)

[6] Navigatsionnyi otchet za 1939 god [A navigation account for 1939 (in Russ.)] (1939), Museum of the port of Dudinka (no number)

[7] The materials of the expedition "Gde "Temza" vstrechaetsia s Eniseem: pamiatniki istorii poliarnogo sudokhadstva na Enisee" [Where Thames meets the Yenisei: historical sites of polar shipping on the Yenisei] (2016).

[8] Hooper J J, Foecke T, Graham L, Weihs T P 2003 Metallurgical analysis of wrought iron from the RMS Titanic *Marine Technology* vol **40** no. 2 pp 73–81

[9] Nelson L H 1968 Nail chronology as an aid to dating old buildings *Technical Leaflet* no. **48** AASLH; 1ed. (12 p.)

[10] Dupen B 2014 Applied Strength of Materials for Engineering Technology. Indiana University – Purdue University Fort Wayne

IOP Conf. Series: Materials Science and Engineering **255** (2017) 012004 doi:10.1088/1757-899X/255/1/012004

Improved model for detection of homogeneous production batches of electronic components

L A Kazakovtsev[1,2], V I Orlov[1], D V Stashkov[1], A N Antamoshkin[1,2] and I S Masich[1,3]
[1] Reshetnev Siberian State University of Science and Technology, 31 Krasnoyarskiy Rabochiy av., Krasnoyarsk, 660037, Russia
[2] Siberian Federal University, 79 Svobodny av., Krasnoyarsk, 660041, Russia
[3] Krasnoyarsk State Agrarian University, 90 Mira av., Krasnoyarsk, 660049, Russia

E-mail: levk@bk.ru

Abstract. Supplying the electronic units of the complex technical systems with electronic devices of the proper quality is one of the most important problems for increasing the whole system reliability. Moreover, for reaching the highest reliability of an electronic unit, the electronic devices of the same type must have equal characteristics which assure their coherent operation. The highest homogeneity of the characteristics is reached if the electronic devices are manufactured as a single production batch. Moreover, each production batch must contain homogeneous raw materials. In this paper, we propose an improved model for detecting the homogeneous production batches of shipped lot of electronic components based on implementing the kurtosis criterion for the results of non-destructive testing performed for each lot of electronic devices used in the space industry.

1. Introduction

The most important units are integrated from electronic devices manufactured as special production batches, which meet special quality requirements [1, 2]. Each device in each of the shipped batches is checked via many destructive and nondestructive tests [1, 3]. Data of the non-destructive testing can be used for analyzing the homogeneity of the shipped lot [3]. For splitting the lot of electronic devices into several assumed production batches, the k-means method [4, 5, 6] or fuzzy clustering methods [7] can be used.

American and European EEE component manufacturers produce electronic devices of special quality classes, Military and Space [8, 9]. Russian manufacturers do not form a special class of components for use in space systems [1, 2].

Some small part of each of the shipped lots of electronic devices is checked via destructive tests. If the homogeneous production batches have not been partitioned, the results of the destructive tests of small number of selected devices cannot be extended to the whole shipped lot.

Quality requirements of the electronic devices in space systems are so high that the range of characteristics measured via quality check tests is very narrow and the quality class and assumed production batch of each device in the lot must be determined via analysis of difference (distance) of test result data vectors which slightly exceeds the precision of the

IOP Conf. Series: Materials Science and Engineering 255 (2017) 012004 doi:10.1088/1757-899X/255/1/012004

measurement tools. Thus, results of each measurement form a finite (discrete) set of possible values defined by the measurement tool precision.

2. The model of production batch as a mixture of Gaussian distributions

Results of tests carried out during the incoming control, additional burn-in testing and other forms of non-destructive testing are represented in a form of data vectors of high dimensionality, up to several hundred dimensions.

If we have test results of a production batch of electronic devices proved to be produced as a homogeneous production batch from a single batch of raw materials, we can estimate the nature of probability distribution [10] of each of measurements (dimension, parameters) of electronic devices in this batch.

Some results in assessment of the nature of distributions of these parameters are listed in [11]. It is shown [11] that the hit frequency histograms of drift values of parameter values obtained from the measurements performed before and after the electrical and thermal trainings have a form close to normal distribution. Also, values of some parameters seem to be normally distributed.

Results of investigating of probability distributions of parameters of the homogeneous production batches of the integrated circuits 1526TL1 are listed below (see table 1). We investigated devices of three production batches of 199, 535 and 98 pieces. All measurements received during non-destructive tests are subjected to a research. Values of each measurement were grouped in 7 intervals.

Table 1. Results of a research of probability distributions of measurements (parameters) of the IC 1526TL1 (production batch of 199 pieces)

Para meter	Avg.	Standard deviation	Kurtosis criterion	Value intervals			
				Interval (from/to)		Hit frequency	Theoretical hit frequency if the probability distribution is Gaussian
T2	-0.03464	0.000355	-0.669	-∞	-0.0353	2	12.1112
				-0.0353	-0.03509	27	22.24904
				-0.03509	-0.03487	34	38.67464
				-0.03487	-0.03466	45	47.19217
				-0.03466	-0.03444	34	40.42747
				-0.03444	-0.03423	23	24.31193
				-0.03423	-0.03401	22	10.26153
				-0.03401	+∞	12	3.772006
				-0.0353	-0.03509	25	22.21356
				-0.03509	-0.03487	36	38.97081
				-0.03487	-0.03466	48	47.66004
				-0.03466	-0.03444	31	40.63498
				-0.03444	-0.03423	26	24.15167
				-0.03423	-0.03401	19	10.00465
				-0.03401	+∞	13	3.557402
T5		0.000351	-0.62321	-∞	-0.0353	1	11.88415
				-0.0353	-0.03509	27	22.35092
				-0.03509	-0.03487	32	39.14586
				-0.03487	-0.03466	49	47.74116
				-0.03466	-0.03444	32	40.54652
				-0.03444	-0.03423	27	23.97934
				-0.03423	-0.03401	18	9.87299
	-0.03465			-0.03401	+∞	13	3.479068

IOP Conf. Series: Materials Science and Engineering **255** (2017) 012004 doi:10.1088/1757-899X/255/1/012004

T12		1E-07	96.95	-∞	-1E-06	2	4.08E-18
				-1E-06	-8.6E-07	0	8.66E-13
				-8.6E-07	-7.1E-07	0	2.47E-08
				-7.1E-07	-5.7E-07	0	9.55E-05
				-5.7E-07	-4.3E-07	0	0.051495
				-4.3E-07	-2.9E-07	0	4.038977
				-2.9E-07	-1.4E-07	0	49.57459
				-1.4E-07	0	197	145.3348
				-1E-06	-8.6E-07	0	5.25E-06
				-8.6E-07	-7.1E-07	0	0.000949
				-7.1E-07	-5.7E-07	0	0.063484
				-5.7E-07	-4.3E-07	0	1.587031
				-4.3E-07	-2.9E-07	0	15.01864
				-2.9E-07	-1.4E-07	0	54.49758
	-1E-08			-1.4E-07	0	195	127.8323
T21		2.93E-07	17.52	-∞	-2E-06	1	9.77E-08
				-2E-06	-1.6E-06	0	0.000331
				-1.6E-06	-1.1E-06	0	0.14289
				-1.1E-06	-7.1E-07	5	8.288913
				-7.1E-07	-2.9E-07	0	70.58328
				-2.9E-07	1.43E-07	185	97.15819
				1.43E-07	5.71E-07	0	22.0597
				5.71E-07	0	8	0.766687
				-∞	2.274286	8	12.57832
				2.274286	2.378571	15	34.21205
				2.378571	2.482857	55	55.05061
				2.482857	2.587143	55	52.43781
				2.587143	2.691429	38	29.56636
				2.691429	2.795714	20	9.860443
	5.03E-09			2.795714	+∞	7	2.184191
T76		0.306458	-0.6699	-∞	-8.17	1	6.834989
				-8.17	-7.96286	15	18.2802
				-7.96286	-7.75571	31	38.50211
				-7.75571	-7.54857	40	52.22657
				-7.54857	-7.34143	49	45.6357
				-7.34143	-7.13429	38	25.68493
				-7.13429	-6.92714	21	9.30728
	-7.50854			-6.92714	+∞	4	2.528227
T143		0.000829	0.1044	-∞	0.0122	1	5.575501
				0.0122	0.012829	12	19.21635
				0.012829	0.013457	34	44.21939
				0.013457	0.014086	46	58.78908
				0.014086	0.014714	64	45.1765
				0.014714	0.015343	31	20.05787
				0.015343	0.015971	5	5.139224
	0.014098			0.015971	+∞	6	0.826082
T154		0.011065	-0.692	-∞	4.73	12	7.381403
				4.73	4.735714	0	12.95928
				4.735714	4.741429	39	24.6027
				4.741429	4.747143	0	35.98084
				4.747143	4.752857	60	40.53894
				4.752857	4.758571	0	35.18796
				4.758571	4.764286	61	23.53026
	4.752613			4.764286	+∞	27	18.81862

IOP Conf. Series: Materials Science and Engineering **255** (2017) 012004 doi:10.1088/1757-899X/255/1/012004

Graphically, statistical results provided in table 1 together and results of data handling of the second production batch (535 pieces) are provided in figures1-3.

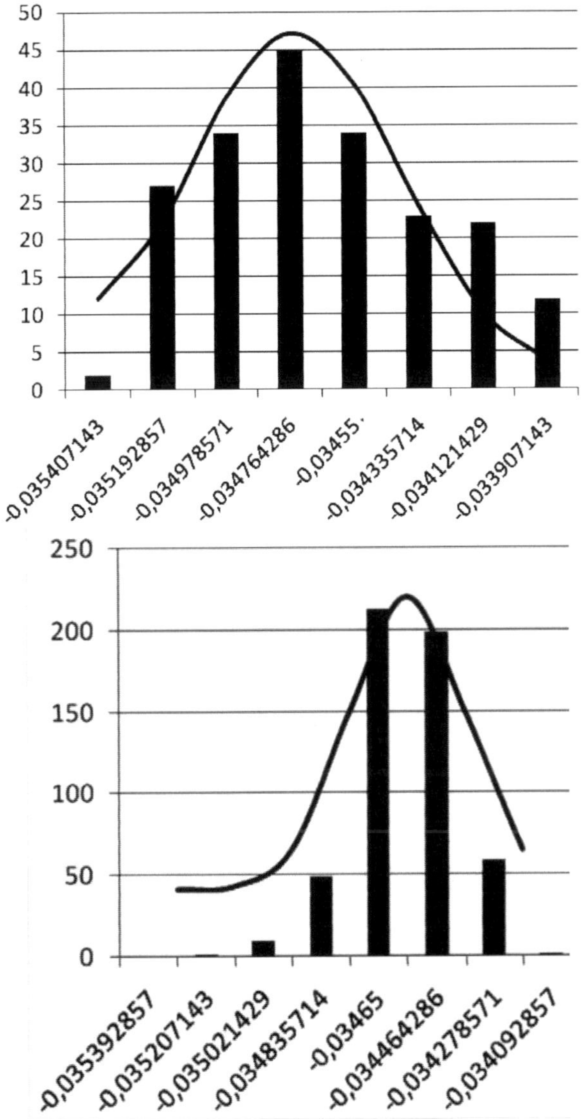

Figure1. Hit frequency histogram of the IC 1526TL1 T2 parameter. The batch No. 1 (199 pieces) above, the batch No. 2 (535 pieces) below

IOP Conf. Series: Materials Science and Engineering **255** (2017) 012004 doi:10.1088/1757-899X/255/1/012004

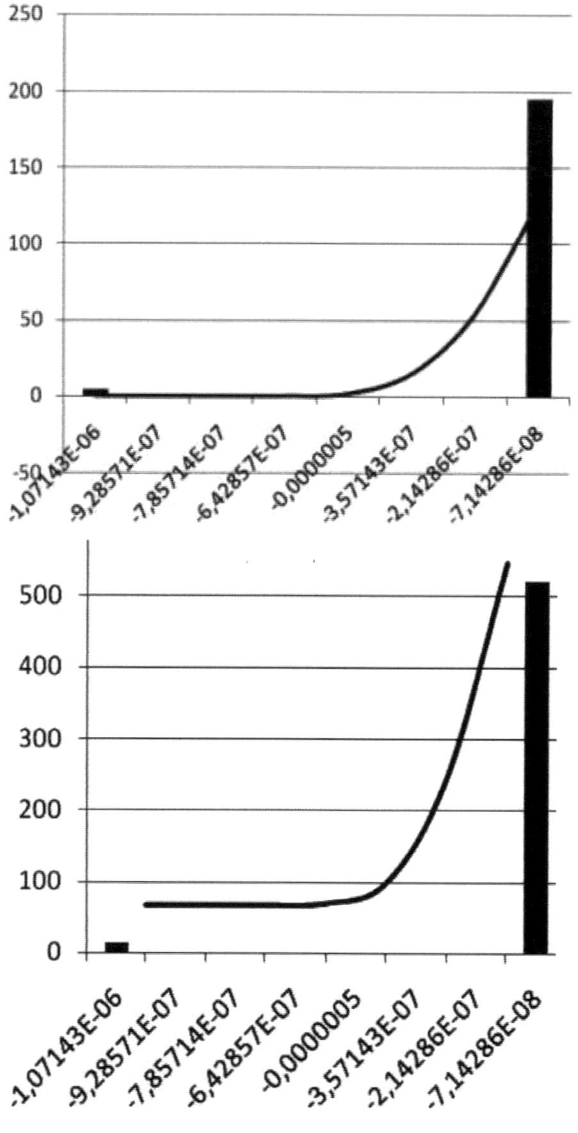

Figure 2. Hit frequency histogram of the IC 1526TL1 T13 parameter. The batch No. 1 (199 pieces) above, the batch No. 2 (535 pieces) below

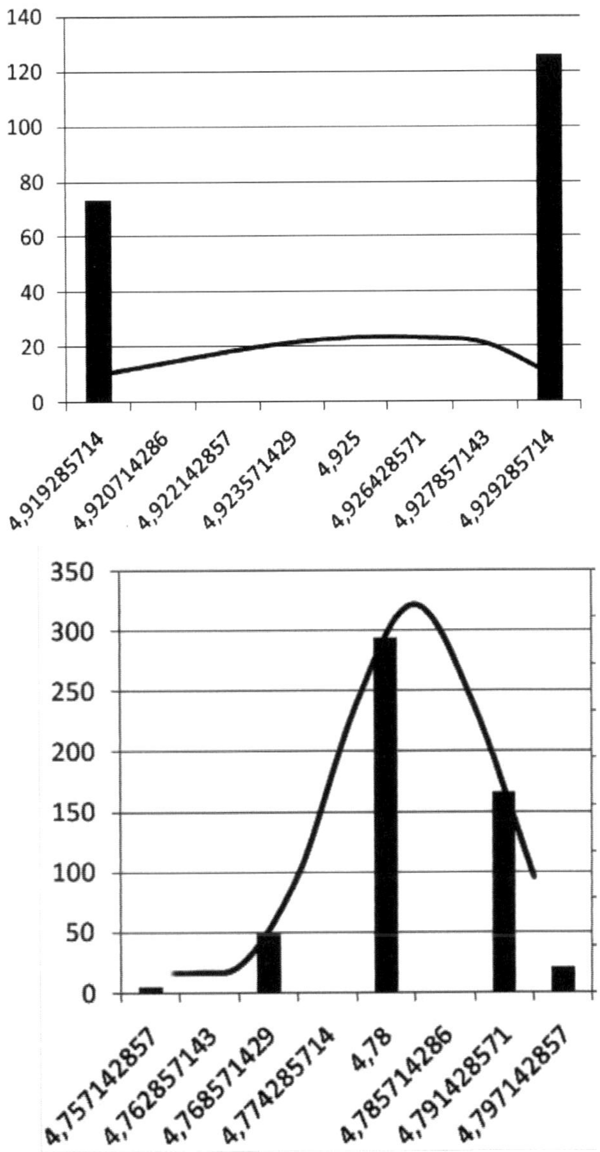

Figure 3. Hit frequency histogram of the IC 1526TL1 T154 parameter. The batch No. 1 (199 pieces) above, the batch No. 2 (535 pieces) below

Analysis of the hit frequency histograms shows that the nature of distributions of parameters in different batches is identical, mean standard deviations are commensurable.

All measurements (parameters, dimensions) of nondestructive tests can be divided into three groups:

A) A group of parameters for which normal the nature of distribution is evident. For the integrated circuit 1526TL1, it is possible to carry T2-T5 (figure 1), T76 to those.

B) Parameters for which the normal nature of probability distribution is visible on the hit frequency histogram, however, the accuracy of measuring instruments is such low that the possible values of this measurement forms a small finite set of values. In this case, the hit frequency histogram contains essential gaps regarding the actual frequency of hit of values in intervals. Please note that assessment of normality of the probability distribution by Pearson's criterion (Chi-square) [10] gives negative answer (hypothesis of normal distribution seems to be wrong). For example, it is possible to carry to such parameters T154 (figure2).

C) Parameters for which the hit frequency histogram has the appearance not corresponding to normal distribution (for example, T12, T13, T21, see figure 3). In certain cases, it is possible to make the assumption of a probability distribution similar with the exponential distribution.

Apparently from table 1, the kurtosis criterion [12]

$$C_{ex} = \frac{\sum_{i=1}^{N}(x_i - \frac{1}{N}\sum_{j=1}^{N} x_j)^4}{(\sum_{i=1}^{N}(x_i - \frac{1}{N}\sum_{j=1}^{N} x_j)^2)^2} - 3 = \frac{\sum_{i=1}^{N}(x_i - \overline{x})^4}{\sigma^4}$$

allows to separate group B parameters from remaining. Here, $x_1 \ldots x_N$ are values of measurements of some parameter of products of a batch from N pieces, x is the average value of this parameter, σ is standard deviation. For such parameters, the values of this criterion are high (more than 10). For normally distributed random variable, this criterion has zero expectation.

Let us note that for different products, parameters of group A form the very considerable part of the whole set of parameters, from 40% to 70%. For parameters of this group and for parameters of group B, admissible boundaries of drift of parameters are set [13].

Model of a combined batch of electronic devices in the form of a mix of spherical or uncorrelated Gaussian distributions [7, 14] allows to separate components of combined batches. At the same time the percent of errors is rather high.

The model of division of a combined batch of electronic devices based on the k-means model [15-17] is also well investigated. At the same time is shown that data normalization with use of the admissible boundaries of parameter drift [13] allows to lower percent of errors in case of division of a combined batch.

Having separated parameters which normality of distribution is not confirmed with use of the kurtosis criterion parameters, it is possible to create a new shortened dataset which parameters have normal distribution with which normality of distribution isn't confirmed. In this case, it is possible to make the reasonable assumption that use of this new dataset allows to increase the adequacy of a model based on a mixture of Gaussian distributions and lower errors ratio.

3. Verification of adequacy of the model

To check of adequacy of the model of division of a combined batch of electronic devices, we used two combined batches as a test example, the batch of integrated circuits 1526IE10 of 870 pieces produced as four homogeneous and a combined batch of amplifiers 140UD25AVK of 53 of pieces [18]. The results for integrated circuits 1526IE10 are presented in Table 2.

4. Conclusion

The nature of probability distributions of separate parameters of electronics devices allows us to conclude that a considerable part of these parameters forms a multidimensional Gaussian distribution. Elimination of the parameters having a distribution other than normal allows to increase adequacy of the model of homogeneous production batches detection based on a mixture of Gaussian distributions.

Table 2. Comparative results of dividing of a combined batch of the IC 1526IE10 (870 pieces)

Model	Batch No 1		Batch No 2		Batch No 3		Batch No 4		Errors ratio
	Quantity according to the model	Quantity of errors in this actual production batch	Quantity according to the model	Quantity of errors in this actual production batch	Quantity according to the model	Quantity of errors in this actual production batch	Quantity according to the model	Quantity of errors in this actual production batch	
In fact	146	-	229	-	424	-	71	-	-
Uncorrelated Gaussians with original dataset with all parameters	244	25	196	41	358	78	72	1	16.7 %
Uncorrelated Gaussians with the new dataset with parameter filtered with kurtosis criterion	125	26	221	8	414	10	110	0	5.05 %
Spherical Gaussians [7]	251	0	189	40	359	76	71	0	13.3 %

Acknowledgments

Results were obtained in the framework of the state task No. 2.5527.2017/8.9 of the Ministry of Education and Science of the Russian Federation.

References

[1] Fedosov V V and Orlov V I 2011 Minimal necessary extent of examination of microelectronic products at inspection test stage *Izvestiya Vuzov. Priborostroenie* **54(4)** 68-62

[2] Kharchenko V S and Yurchenko Yu B 2003 Rating of fault-tolerant onboard complexes frames at usage electronic components industry *Tekhnologiya i konstruirovanie v elektronnoy apparature* **2** 3–10

[3] Kazakovtsev L A, Orlov V I, Stupina A A and Masich I S 2014 Zadacha klassifikatsii elektronnykh komponentov [Problem of electronic components classifying] *Vestnik SibGAU* **4(56)** 55-61

[4] Ackermann M R et al 2012 StreamKM: A Clustering Algorithm for Data Streams *J. Exp. Algorithmics* **17** 2.4:2.1-2.30

[5] Kanungo T, Mount D M, Netanyahu N S, Piatko C D, Silverman R and Wu A Y 1999 Computing nearest neighbors for moving points and applications to clustering *Proc. of the tenth annual ACM-SIAM symp. on Discrete algorithms (Society for Industrial and Applied Mathematics)* 931-932

[6] Kazakovtsev L A, Stupina A A and Orlov V I 2014 Modification of the genetic algorithm with greedy heuristic for continuous location and classification problems *Sistemy upravleniya i informatsionnye tekhnologii* **2(56)** 31-34
[7] Orlov V I, Stashkov DV, Kazakovtsev LA andStupina AA2016 Fuzzy clustering of EEE components for space industry *IOP Conference Series: Materials Science and Engineering* **155** P 012026
[8] Hamiter L 1991 The History of Space Quality EEE Parts in the United States ESA *Electronic Components Conf. ESTEC* (Noordwijk, The Netherlands, 12–16 Nov 1990 ESA SP-313)
[9] Kirkconnell C S et al 2014 High Efficiency Digital Cooler Electronics for Aerospace Applications *Proc. SPIE 9070, Infrared Technology and Applications* XL 90702Q (June 24, 2014)
[10] Ader H J 2008 Phases and initial steps in data analysis *Advising on Research Methods: A consultant's companion* Ed H J Ader and G J Mellenbergh (Huizen, the Netherlands: Johannes van Kessel Publishing) 333–356
[11] Fedosov V V 2015 *Voprosy obespecheniya rabotosposobnosti elektronnoy komponentnoy bazy kosmicheskikh apparatov* [Questions of operability support of an electronic component basis in the spacecraft equipment] (Krasnoyarsk: Siberian State Aerospace University)
[12] Hazewinkel M 2001 Excess coefficient *Encyclopedia of Mathematics* (Springer)
[13] Fedosov V V, Kazakovtsev L A and Masich I S 2016 Metod normirovki ishodnyh dannyh ispytanij elektroradioizdelij kosmicheskogo primenenija dlja algoritma avtomaticheskoj gruppirovki [Normalization method for initial testdata of electrical radio products in space application for automatic grouping algorithm] *Sistemy upravleniya i informatsionnye tekhnologii* **65(3)** 92-96
[14] Stashkov D V 2017 Algoritmy poiska s cheredujushhimisja okrestnostjami dlja zadachi razdelenijas mesi raspredelenij [Variable neighborhood search algorithms for problem of separating of a mix of distributions] *Sistemy upravleniya i informatsionnye tekhnologii* **1** 18-24
[15] Kazakovtsev L A, Orlov V I and Stupina A A 2015 Vybor metriki pri klassifikacii elektroradioizdelij po proizvodstvennym partijam [On distance metric for the system of automatic classification of the EEE devices by production batches] *Programmnye produkty i sistemy* **2** 124–129
[16] Kazakovtsev L A, Antamoshkin A N and Masich I S 2015 Fast Deterministic Algorithm for EEE Components Classification *IOP Conf. Series: Materials Science and Engineering* **94** article ID 012015
[17] Stupina A A and Kazakovtsev L A 2015 Fast genetic algorithm with greedy heuristic for p-median and k-means problems *2014 6th International Congress on Ultra Modern Telecommunications and Control Systems and Workshops (ICUMT 2014)* 602-606
[18] Orlov V I, Kazakovtsev L A and Masich I S 2016 Silhouette Criterion for Automatic Grouping Algorithm of Spaceship Electronic Components *Vestnik SibGAU* **17 (4)** 883-890

Analysis and improvement of calculation procedure of high-speed centrifugal pumps

E M Kraeva[1] and I S Masich[1,2]
Reshetnev Siberian State University of Science and Technology,
31 Krasnoyarskiy Rabochiy av., Krasnoyarsk, 660037, Russia
[2] Krasnoyarsk State Agrarian University, 90 Mira av., Krasnoyarsk, 660049, Russia

E-mail: em-kraeva@yandex.ru

Abstract
The model of flow around a flowing part of a high-speed centrifugal pump with a semi-open impeller is presented. The calculated ratios for design of low-flow pumps are obtained and confirmed experimentally.

1. Introduction

The development of aerospace technology, expansion of the range of problems solved with the help of power plants of aircraft put forward for researchers the problem of increase of efficiency of supply pumping systems with low consumption, their design excellence and forecasting energy parameters with a given level of stability. The solution to this problem contributes to the quality of design, speeding up processing and commissioning of modern power plants of aircraft [1].

High-speed centrifugal pumps (HSCP) at an angular speed of the rotor up to 10 000 rad / s is widely used as a part of a turbopump assembly of liquid rocket thrusters and power plants of aircrafts, which is caused by a wide range of changes in their regime parameters. For example, when the angular velocity is from 3000 to 10000 rad / s, the value $\dot{V}/\omega$ reaches at least 10^{-7} m^3 subject to $\mathrm{Re}_\omega > 10^5$. Reduction in the supply $\dot{V}$ in such pumps, along with an increase in the angular velocity of the rotor, usually results in a decrease $\dot{V}/\omega$ less than the value $(\dot{V}/\omega) = 1 \cdot 10^{-6}$ m^3, which is the maximum permissible for centrifugal pumps with closed impeller [1]. Therefore, HSCP with an open and semi-open impeller are widely used.

2. Visualization of the flow pattern in an impeller

For different ratios of width and height of the channel $b/h = 3,2,1,0.5$ M. Van Dyke has been presented visualization pictures of flow over a rectangular cavity [2]. By decreasing the width of cavity under primary vortex, the secondary one starts to grow. When the ratio of width to height is tend to zero, an infinite sequence of vortices is formed, each vortex is weaker than its predecessor. In order to clarify the main features of hydrodynamics in semi-open channels of impellers, series of experiments are carried out, including experiments on flow visualization in such impellers.

Tests over a wide range of regime pump parameters ($\dot{V} = 0...1.5\dot{V}_{nom}$) and change of the axial clearance from 0.5 to 3 mm revealed that for different radii the constancy of the ratio of the radial component of velocity υ_R to the circumferential velocity is hold: $\upsilon_R/U = \varphi_R$. This gives grounds to express the speed of radial flow in an axial gap from the periphery of the impeller to the circumferential direction with respect to the impeller in the form

$$\upsilon_R = U\varphi_R = UK_R\varphi_\Lambda, \quad (1)$$

where K_R is the experimental constant, $K_R = 0.45$.

Consequently, as at the entrance of the impeller with semi-open blades and on its radius, not all of the flow channels of the impeller is filled with the stream moving from the entrance of the impeller to the exit. Partially its flowing part is filled by backflow. Swirling in the direction of rotation of the impeller, reverse currents flow into the adjacent channel to a smaller radius and are addicted to the back of the flow (figure 1). This creates a vortex zone in which the liquid is not involved in the supply flow through the pump. With increasing supply through the pump with $\omega = const$ parameters of the vortex zone are reduced, resulting a decrease in the flow of the radial twist.

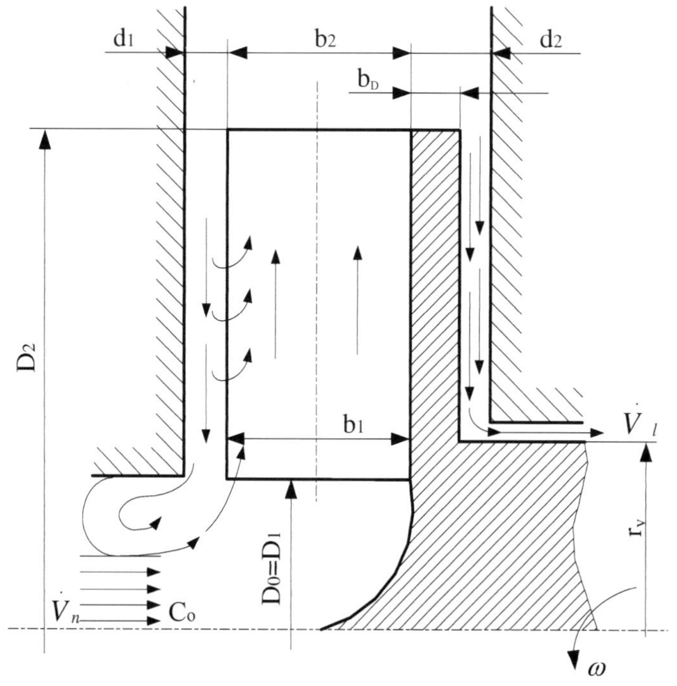

Figure 1. Scheme of the turbulent flow of the stream in the channel under the flow around the blade

Based on the studies on stream visualization [3] and measuring the hydrodynamic parameters of flow in the gap between the rotating impeller with open ends of the blades and smooth hull, a model of the jet-vortex flow around channels of an impeller can be represented (figure 2). The liquid flow in the inter-blade channel of the impeller is exposed to direct force influence of the blades. The liquid in the axial gap a_1 is twisted due to friction forces and slips relative to ends of the blades of the impeller. Thus, on the same radius the fluid particles in the channel and the axial gap a_1 move with different circumferential speed, which leads to their relative movement in the radial and axial directions.

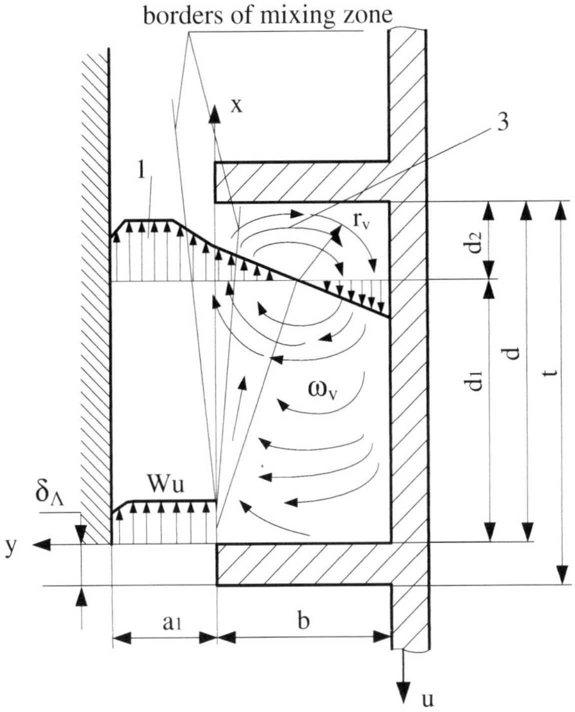

Figure 2. Scheme of interaction of the incoming stream of axial clearance and the stream in the pump impeller channels:
1 – the zone of recirculating flow in the impeller channel;
2 – the zone of incoming flow of axial clearance;
3 – the borders of the mixing zone.

In the circumferential direction of fluid flow in the gap a_1 relative to the blades causes the formation circulation zone in channels of the impeller as a result of the stream flows around cavities. It is known that the fluid flow in the channel behind a poorly streamlined body (in our case these are the blades of HSCP) having separated flows, which are characterized by the formation of reverse currents and vortices. This process is largely determined by the ratio of blade width b to the distance between them $d = t - \delta_\Lambda$, the boundary layer thickness on the wall in front of the channel and a relative height of the channel. The interaction between the jet and the fluid in the channels of the impeller leads to the appearance of the circulation flow in the channel behind the blade. In the cross section under $y = 0$ the stream due to its compression in the axial gap a_1 is uniform, and as a result of automodelity of turbulent jets [4], the parameters of the circulation zone do not depend on the number Re. Then for the mixing zone 2 the value of relative velocity of the flow is expressed as:

IOP Conf. Series: Materials Science and Engineering **255** (2017) 012005 doi:10.1088/1757-899X/255/1/012005

$$W_U = (U - U_l) = U(1 - \overline{\omega}_l) \quad (2)$$

3. The boundaries for the different zones of flow

Using the basic regularities for free flat jets [4], based on the known experiments on strain of any irregular profile to a jet one at a very small distance from the blade edge, we obtain bounds for the different flow zones under the flow in an interblade channel of finite length.

For symmetric about the axis X boundary of the mixing zone 2 dividing the flow in the channel from the stream in the gap a_1, its position can be determined by the expression [4]

$$y = \pm 0.08829x. \quad (3)$$

Comparing the obtained pressure distributions based on visualization of flows in the fixed cavity and in rotary channels of the semi-open impeller gives grounds to assume that in the channels there is an intense vortex flow with the axis of vortex directed along the radius of the impeller and with displacement of the vortex to the pressure side of the blade. The circulation area in the channel of the semi-open impeller (figure 3) let us roughly divide into two areas: the area of intensive rotational motion along a circle with radius r_v and the area of circulation along the trajectories of the elliptical nature. The size of vortex zone radius on base of equation (3) equal to

$$r_v = 0.519(b - 0.08829d). \quad (4)$$

One of the characteristics of the vortex zone is the ratio of the circumferential velocity and the velocity of the incident flow W_u at the external boundary of the vortex zone

$$\overline{W}_v = \frac{W_v}{W_u} = f\left(\frac{b}{a}\right).$$

Analysis of special experiments [5, 6, 7] on determining the estimated value $\overline{W}_v$ gives the basis to take it for our investigations equal to 0.3. The same value $\overline{W}_v$ is shown in a number of works for the fixed channel. Then the expression for the angular speed of rotation of the vortex in channels of the impeller on the radius R, with taking into account the value $\overline{W}_v = 0.3$ and the dependence (2), can be written as

$$\omega_v = \frac{\overline{W}_v(1 - \overline{\omega}_l)\omega R}{r_v} = \frac{\overline{W}_v(1 - \overline{\omega}_l)\omega R}{0.519(b - 0.08829d)}. \quad (5)$$

Experimental data generally confirm the adopted model for calculating channels of the semi-open impeller in the form of three-zone jet-vortex flow, which allows to calculate its geometric parameters.

The parameters of the mixing and vortex flow zone have decisive influence on the sizes of the impeller channel. On the basis of the experiments of works [2, 3, 4], it should be noted that for flow around the square channel $b/d = 1$ there is one stable vortex rotating almost like a solid body. With further increase in the channel depth to $b/d = 2$, there are two vortex cells observed that arranged one above the other and having opposite directions of rotation. In order to minimize hydraulic losses in the impeller channels, a multivortex flow is unacceptable both in depth and in the width of the channel. Proceeding from this condition and assuming that the average velocity at the end of the section d_1 for the forward and reverse flows of circulating zone are equal (figure 2), according to the decision G.N. Abramovich [4], for the area in which energy of the reverse single-vortex flow reaches a maximum, we get

$$d_1 = (4.75 \div 5.2)b\,. \tag{6}$$

Similarly, we can get the size of the second section with length d_2 from the terms of the approximate equality of the averaged values of the energy in the forward and reverse flows in cross section $x = d_1$:

$$d_2 = (0.980 \div 0.852)b\,. \tag{7}$$

Thus, the total length of the deep channel for single-vortex flow is

$$d = d_1 + d_2 = (6.18 \div 6.05)b\,. \tag{8}$$

The flow in the shallow channel is characterized by an elongated boundary of the mixing zone and its accession to the surface of the channel at the point $x = d_1$. Taking into account the relation (3) we have: $d_1 = 11.36b$. Then, taking into account that the vortex flow on the pressure side of the blade is formed by the jet overflowing to backside, we find the greatest value of the channel values:

$$d_{max} = (11.36 + 1.14)b = 12.5b\,. \tag{9}$$

From (4) the minimum possible value of the channel is

$$d_{min} = 1.1423b\,. \tag{10}$$

The minimum width blade at the outlet of the impeller can be finding from the condition of conservation of the stream core in the inter-blade channel that will at the same time the condition of applicability the theory to the calculation of impeller channels [4].

The performed experimental research, analysis and visualization of streamlines data obtained fully confirmed the calculated ratio of the sizes of interblade impeller channels that are expressed in the form of dependencies for the entry parameters on the blade with thickness $\delta_{1\Lambda}$ and the channel angle $\beta_{1\Lambda}$:

$$\left(\frac{\pi D_1}{z} - \frac{\delta_{1\Lambda}}{\sin \beta_{1\Lambda}} \right) \geq K_1 b_1\,, \tag{11}$$

where K_1 is the coefficient of impeller channel width at the entrance, $K_1 = 1.14$.

For the output from the impeller we similarly write

$$\left(\frac{\pi D_2}{z}-\frac{\delta_{2\Lambda}}{\sin\beta_{2\Lambda}}\right)\leq K_2 b_2, \tag{12}$$

where K_2 is the coefficient of impeller channel width at the output, $K_2 = 6.18...12.5$.

Value for impeller channel width at the output taking into account density τ_2 of the profiles lattice is

$$b_2 = \frac{(1-\overline{D}_1)D_2}{K_2\tau_2\sin\left(\frac{\beta_{1\Lambda}+\beta_{2\Lambda}}{2}\right)}-\frac{\delta_{2\Lambda}}{K_2\sin\beta_{2\Lambda}} \tag{13}$$

Similarly, with regard to (12) for the impeller parameters at the density lattice τ_1 we obtain:

$$b_1 = \frac{(1-\overline{D}_1)D_1}{K_1\tau_1\sin\left(\frac{\beta_{1\Lambda}+\beta_{2\Lambda}}{2}\right)}-\frac{\delta_{1\Lambda}}{K_1\sin\beta_{1\Lambda}} \tag{14}$$

Dependencies for two values of channel width coefficient $K_2 = 12.5$ and 6.18 calculated by the expression (13), in which limits all the experimental points of HSCP with the half-open impeller lie, are shown in figure 3. It should be noted that in the finishing process the number of HSCP to achieve acceptable parameters on the pressure and the efficiency, it was sufficient to increase the blade width to the value approaching the upper curve ($K_2 = 6.18$).

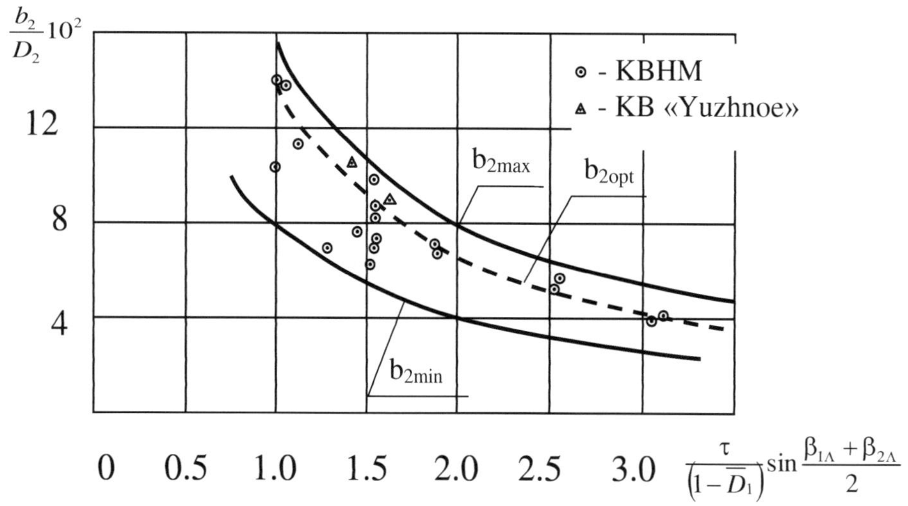

Figure 3. The calculated dependences on the change in blade width of the semi-open impeller ▲ ⊙ – experimental points based on the results of tests of centrifugal pumps with a semi-open impeller of the various design bureaus

4. Improving the energy transfer processes

One of the real ways to increase the specific power of the pump consists in improving the processes of energy transfer to fluid flow in the flowing part of the pump, which is realized by setting in its cavities the equalization elements in the form of perforated lattices and grids. It should be noted that the equalization elements are widely used in the stabilization of flows in immobile channels. The equalization elements are detailed study on their interactions on the flow in the part of alignment of velocity field and pressure and eliminating fluctuations in the pipelines.

For pumps with a specific supply $\dot{V}/\omega = (1...5)\cdot 10^{-7}$ m^3, the optimal parameters of the grid with the flow cross section coefficient $\bar{f} = 0.4...0.56$ are calculated on the present procedure. The results of comparative tests of pumps with the impellers without the grid and with the grid ($\bar{f} = 0.5$) on the output are shown in figure 4. Due to the alignment of the stream structure the pump pressure is increased by 15 ... 20%. At high speed pumps, nonuniformity is higher and such construction provides increased pressure up to 30%. The presence of the grid at the entrance of the impeller leads to improved cavitation characteristic of the pump. Installation of equalization elements reduces unevenness of fluid velocity field behind the impeller, reduces deviation of the fluid flow from the direction of the blades at the outlet of the impeller and pressure pulsations.

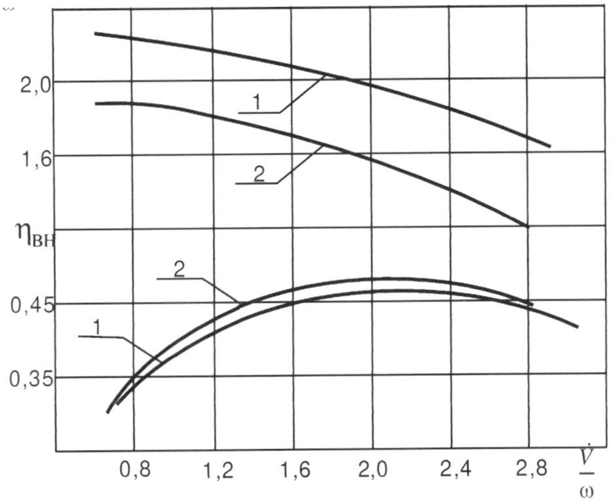

Figure 4. Energy characteristics of the pump:
1 – the impeller with grid at the output ($\bar{f} = 0.5$);
2 – the impeller without grid

5. Conclusion

Further experimental study of HSCP with the semi-open impeller in a wide range of constructive and regime parameters, as well as analysis of carried out constructions of HSCP received as a result of their repeated improvements (over twenty sizes) have shown the acceptability of dependencies (12) and (13) taking into account lattice density for calculation basic structural relations of HSCP with the impeller of the semi-open type.

In order to generalize the studies carried out, recommendations have been developed for calculating the energy parameters of the HSCP with impellers of semi-open types.

Experimental data of the authors and other researchers confirmed the main theoretical positions. The results of the studies are presented in the form of generalized dependencies, which make it possible to widely use them in engineering practice.

Acknowledgments

Results were obtained in the framework of the state task No. 2.5527.2017/8.9 of the Ministry of Education and Science of the Russian Federation.

References

[1] Kraev M V, Lukin V A and Ovsyannikov B V 1985 *Low-flow pumps of aviation and space systems* (Moscow: Mashinostroenie) p 128
[2] Van Dyke M 1982 *An Album of Fluid Motion* (Stanford: Parabolic Press) p 176
[3] Kraev M V, Ovchinnikov B V and Shapiro A S 1976 *Hydrodynamic seal high-speed shafts* (Moscow: Mashinostroyeniye) p 104
[4] Abramovich G N 1960 Theory of turbulent jets (Moscow: Fizmatgiz) p 716
[5] Kraev M V and Kraeva E M 2013 Dynamics of cavity development in the centrifugal pump impeller *Russian Aeronautics* **56 (2)** 166-171
[6] Kraev M V and Kraeva E M 2012 Analysis of three-dimensional flow in high-speed pump centrifugal impeller *Russian Aeronautics* **55 (4)** 383-387
[7] Kraeva E M 2012 Optimization of the centrifugal impeller flow passage in high-speed pumps *Russian Aeronautics* **54 (2)** 154-158

Particular Charactristics of the Synthesis of Titanium Nitride Nanopowders in the Plasma of Low Pressure Arc Discharge

A V Ushakov[1,2], I V Karpov[1,2], A A Lepeshev[1,2], L Yu Fedorov[1,2], E A Dorozhkina[3], O N Karpova[3], A A Shaikhadinov[1,2], V G Demin[1,2], E P Bachurina[1,2], D V Lichargin[1], A K Abkaryan[2], G M Zeer[2], S M Zharkov[2,4]
[1]Federal Research Center "Krasnoyarsk Scientific Center" of the Russian Academy of Sciences, Krasnoyarsk, 660036, Russia
[2]Siberian Federal University, Krasnoyarsk, 660041, Russia
[3]Reshetnev Siberian State University of Science and Technology, Krasnoyarsk, 660037, Russia
[4]Kirensky Institute of Physics, Federal Research Center KSC SB RAS, Russia

E-mail: sfu-unesco@mail.ru

Annotation. The method of producing TiN nanopowders in the plasma of low pressure arc discharge in nitrogen atmosphere is discussed. The influence of gas pressure on nanopowder dispersion is studied. It is shown that the particles have nanometer size and a narrow particle size distribution.

Among the methods used to produce nanoscale powders [1-21], especially of refractory metals and their compounds, the most important place is given to the plasma technology, which provides high heating temperatures of a material and steep temperature gradients, so it is very attractive and technically simple method. The presence of supersaturated vapors of the initial compounds in the plasma and their high quench rate causes the formation of nanoparticles.

Recently the interest to the synthesis of nanomaterials (NM) based on refractory compounds, including titanium nitride, has increased. Various heating methods are used for the evaporation of material, such as laser, electron beams, etc.

In this work the arc evaporator, used in the process of TiN synthesis, had the following characteristics [22, 23]: a discharge current of 500 A, an intensity of the longitudinal magnetic field excited by the focusing coil on the surface of the cathode, of 80 Oe. Titanium of technical grade VT 1-00 was used as a cathode. In order to begin plasma chemical reactions, the chamber was preliminary evacuated to a pressure of 1 mPa and then it was filled by gaseous N_2. Evaporation was performed at two pressures of nitrogen: 0.1 Pa (TiN1 sample) and 10 Pa (TiN2 sample). Before evaporation the cathode was heated to 1200 K. Titanium nitride NP was accumulated for 10 minutes. TiN1 sample was of dark brown color, and TiN2 – of black color.

Morphological composition of NP was studied by scanning electron microscope JEOL JSM-7001F and transmission electron microscope JEM-2100 JEOL. Fig. 1 shows a typical picture of TiN1 sample. The powder includes particles of two types: spherical particles with size from 0.5 μm to 3 μm, wherein fine condensate dropped, and spherical particles ranging in size from 0.5 μm to 10 μm, which represent agglomerates of smaller particles of 0.1 μm order. The sample TiN2 for the electron microscopy studies was prepared as follows: the powder

IOP Conf. Series: Materials Science and Engineering **255** (2017) 012006 doi:10.1088/1757-899X/255/1/012006

was placed in isopropyl alcohol and then it was dispersed in an ultrasonic bath for 2 minutes, and a solution drop was then placed on a carbon film substrate, located on an electron-microscopic supporting grid. The carbon film substrate was 10-15 nm thick. Fig. 2 shows a high-resolution electron microscopic image of TiN nanoparticles. The average particle size was 4 nm, some particles were up to 5-6 nm in size. TiN nanoparticles have a crystalline structure, as evidenced by the atomic planes observed in the images. Particle shape was usually nearly spherical.

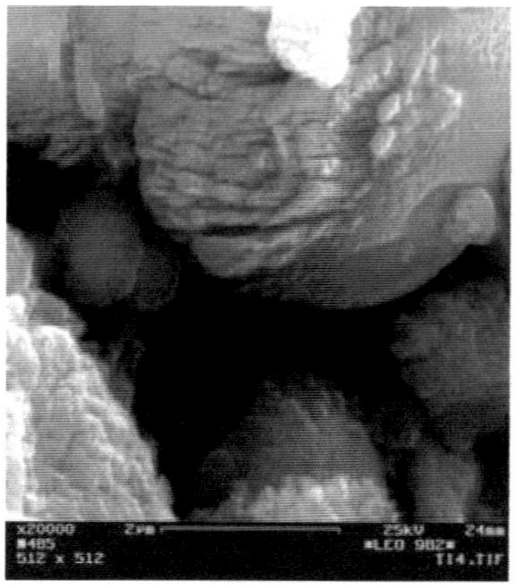

Figure 1. SEM micrographs of TiN1 sample.

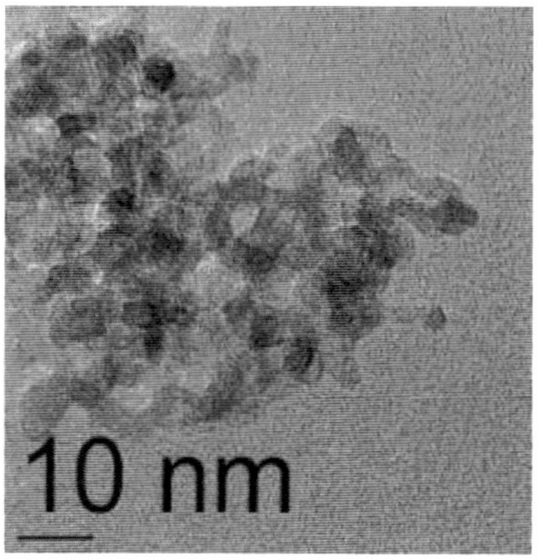

Figure 2. TEM micrographs of TiN2 sample.

X-ray diffraction data were obtained using diffractometer Advance D8. CuKα monochromatic radiation was used. The scanning was carried out at room temperature in the angular range 30°-120° in 2θ with a step of 0.04°.

The X-ray pictures of the samples are shown in Fig. 3. The experimental X-ray picture of TiN2 sample and the estimated, obtained from full-profile adjustment, are given for comparison. The basic phase of the sample TiN1 is a cubic TiN with a lattice parameter a = 0.4242 nm. The lattice parameter is almost identical with the diffraction standards database

ICDD-JCPDS-38-1420. There is a small amount of Ti_2N, the peaks of which are indicated by asterisks in Fig.3.

TiN2 sample represents a titanium nitride cubic phase with a lattice parameter of a = 0.4208 nm, which is substantially less than a lattice parameter of the standard TiN. The particularities of X-ray peaks showed the presence of two fractions in the sample, significantly different in crystallite size. This conclusion was based on the fact that the diffraction maxima have anomalously sharp peaks under the large integrated width. This complex shape could be explained only by a superposition of two peaks, significantly different in width. Therefore, during a full-profile adjustment two fractions with different crystallite size and different composition percentage were introduced to the model. Full-profile adjustment of microstructural parameters revealed that the majority of the sample (~ 98%) is the fraction with an average crystallite size of 8 nm. The remaining 2% of the material consists of coarse fraction with grain size of > 150 nm.

The experimental results clearly indicate sufficient difference between powders obtained at various pressures of gas mixture. The structure of TiN1 sample is like the structure of the coatings obtained by the method of condensation and ion bombardment (CIB). The composition of the generated plasma in the working chamber is not uniform. In particular, upon evaporation from the surface of the titanium cathode, plasma state is formed comprising single, double or triply charged particles. Drop component and fragments of the cathode material may also present in the plasma in addition to the charged particles. The multicomponent character of plasma creates the conditions for the formation of condensate which is non-equilibrium in composition and structure. Fig. 3 shows that the X-ray reflexes are substantially widen. The analysis of widening revealed the presence of micro-tension and nanodispersed structure. Moreover, a greater contribution to line widening is due to defects and tensions in the powder structure. The estimated values of micro-tension in powders TiN1 and TiN2 are 0.8% and 0.6% respectively.

The authors of [24] pointed out that the basic process of nitrides formation occurs mainly on the substrate. Content of a non-metal in the condensate is a linear function of the energy of metal ions which are condensed on the substrate. The energy of the ions has its boundary value, at which a complete binding of the non-metal flow, dripped on the substrate, achieves. However, the work [24] does not consider the energy of electrons from the plasma of arc discharge, which can be considerable. Growth of condensate causes decrease of heat transfer to the substrate, and a determining factor in the process of nitrogen binding becomes a local overheating by the electron current. The other situation is observed in the processes of nitride formation under the pressures which make essential plasma phenomena. The electron microscopy of TiN2 sample shows that there are no large powder particles, such as represent in TiN1 sample. The size characteristics of crystallite TiN2 sample, obtained from full profile analysis of the diffraction patterns, are in good agreement with the results of microscopy and adsorption analysis.

It is known [25], the character and intensity of heat and mass transfer between the particles of the dispersed phase and the plasma flow are determined by energy content of the plasma flow, the physicochemical properties of the fine phase particles, nature of plasma-forming gas and, finally, the character of interaction between the plasma flow and fine dispersed phase, which mainly depends on the Knudsen criterion. When the plasma flow regime becomes molecular (Kn > 1), the convection processes are insignificant. Under these conditions, heat and mass transfer processes are most reasonable viewed from the standpoint of the kinetic theory. In this regard, the estimation of the possibility of evaporation of micro-droplet fraction particles in the plasma of arc discharge can be made due to the following fact: there is a flow of fast electrons in the cathode plasma and they are accelerated in an electrostatic field arising due to the ions which diffuse into the cathode side. The estimations show that the basic flow of heat from the plasma to the particle is due to the energy brought by the ions. The part of the energy brought by the electrons in the considered range of electronic temperature does not exceed 15%. The estimation of the time of complete evaporation of titanium particle with a diameter of 50 microns at an energy of the electrons of about 3.5 eV and a plasma concentration of 10^{21} m^{-3} gave the value of 10^{-5} s, which is much less than the time while a particle of microdroplet fraction exists in the cathode region of the plasma, at a rate of its displacement not exceeding 100 m/s.

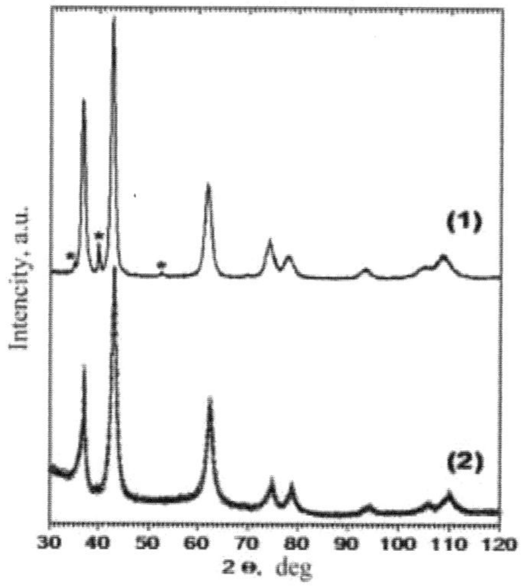

Figure 3. X-Ray diffraction pattern of the TiN1 (*1*) and TiN2 (*2*).

Thus, the estimations show that, in condition of the sufficient energy content of the plasma flow, the process of cathode evaporation in low pressure arc discharge consists of two stages: firstly the cathode material is sprayed in a liquid phase in a cathode spot, and then is completely evaporated in a vapor-plasma flow. Furthermore, due to mixing of a metal vapor with a flow of ionized carrier gas, vapors are overheated, thus preventing premature condensation of the vapors and causes dissociation of already formed clusters. Uniform rate of flow cooling results in a very high size homogeneity of the resulting NM. Cooling the hot mixture by gas-dynamic expansion lets to achieve a rate of flow cooling of up to 10^7 K/s. It can be also assumed that a principal role in the processes of nitrides formation is played by plasma-chemical synthesis, i.e., titanium nitride formation directly in the plasma.

One of the possible mechanisms of droplet breaking is described in [26]. The industrial method for producing NM with additional breaking of particles is developed, its explanation reveals the essence of powders use. The method is based on electric arc spraying of metal in the atmosphere of a diatomic gas. In the powerful DC arc at atmospheric pressure metal droplets are formed, they are quickly saturated by almost completely dissociated gas - hydrogen or nitrogen. Upon cooling, the concentration of dissolved gas tends to the equilibrium one, defined by the Stiverts law. During a gassing process a drop breaks to a more dispersed state.

Thus, a dispersion of titanium nitride NM, produced in the plasma of low pressure arc discharge, depends on the gas pressure in the chamber. Increasing the pressure of the gas mixture leads to extremely high particle size uniformity of the produced NM and sharp decrease in the average size of particles.

Acknowledgements. The reported study was funded by Russian Foundation for Basic Research, project no 17-48-240806, Government of Krasnoyarsk Territory, Krasnoyarsk Region Science and Technology Support Fund to the Agreement №14.

References

[1] Lepeshev A, Sordelet D, Rozhkova E and Ushakov A 2011 *J. of Cluster Sci.* **22** 289. doi: 10.1007/s10876-011-0378-2.
[2] Lepeshev A, Rozhkova E, Karpov I, Ushakov A and Fedorov L 2013 *Phys. of the Solid State* **55** 2531. doi: 10.1134/S1063783413120202.
[3] Ushakov A, Karpov I, Fedorov L and Lepeshev A 2014 *J of Friction and Wear* **35** 7. doi: 10.3103/S1068366614010103.
[4] Ushakov A, Karpov I, Lepeshev A, Petrov M and Fedorov L 2014 *JETP Letters* **99** 99. doi: 10.1134/S002136401402009X.
[5] Ushakov A, Karpov I, Lepeshev A and Petrov M 2015 *J. Appl. Phys.* **118** 023907. http://dx.doi.org/10.1063/1.4926549.
[6] Ushakov A, Karpov I, Fedorov L, Lepeshev A, Shaikhadinov A, and Demin V 2015 *Theor. Foundations of Chemical Engineering* **49** 743.
[7] Fedorov L, Karpov I, Ushakov A and Lepeshev A 2015 *Inorg. Mater.* **51** 25. doi: 10.1134/S0020168515010057.
[8] Lepeshev A, Bayukov O, Rozhkova E, Karpov I, Ushakov A and Fedorov L 2015 *Phys. of the Solid State* **57** 255. doi: 10.1134/S1063783415020249.
[9] Ushakov A, Karpov I, Lepeshev A, Petrov M and Fedorov L 2015 *Phys. of the Solid State* **57** 919. doi: 10.1134/S1063783415050303.
[10] Ushakov A, Karpov I and Lepeshev A 2015 *Phys. of the Solid State* **57** 2320. doi: 10.1134/S1063783415110359.
[11] Ushakov A, Karpov I, Lepeshev A, Fedorov L and Shaikhadinov A 2016 *Technical Phys.* **86**, 103. doi: 10.1134/S1063784216010230.
[12] Ushakov A, Karpov I and Lepeshev A 2016 *Technical Phys.* **86**, 260. doi: 10.1134/S1063784216020262.
[13] Lepeshev A, Karpov I, Ushakov A and Nagibin G 2016 *J. of Alloys and Compounds* **663** 631. doi:10.1016/j.jallcom.2015.12.168.
[14] Lepeshev A, Karpov I, Ushakov A, Fedorov L and Shaikhadinov A 2016 *Intern. J. of Nanoscience* **15** 1550027.
[15] Karpov I, Ushakov A, Lepeshev A and Zharkov S 2016 *Vacuum* **128** 123. doi: 10.1016/j.vacuum.2016.03.025.
[16] Ushakov A, Karpov I, Lepeshev A and Petrov M 2016 *Vacuum* **133** 25. doi: 10.1016/j.vacuum.2016.08.007.
[17] Lepeshev A, Rozhkova E, Karpov I, Ushakov A, Fedorov L, Dorozhkina E and Karpova O 2016 *IOP Conf. Series: Materials Science and Engineering* **155** 012014. doi: 10.1088/1757-899X/155/1/012014.
[18] Karpov I, Ushakov A, Lepeshev A, Dorozhkina E, Karpova O, Shaikhadinov A and Demin V 2016 *IOP Conf. Series: Materials Science and Engineering* **155** 012013. doi: 10.1088/1757-899X/155/1/012013.

[19] Ushakov A, Karpov I and Lepeshev A 2017 *Journal of Superconductivity and Novel Magnetism* **30** 311. doi: 10.1007/s10948-016-3709-6.
[20] Rudenko K, Miakonkih A, Rogojin A, Bogdanov S, Sidorov V and Zelenkov P 2016 *IOP Conf. Ser.: Mater. Sci. and Engineering* **122** 012029. doi: 10.1088/1757-899X/122/1/012029.
[21] Bogdanov S, Lelekov E, Kovalev I, Zelenkov P and Lelekov A 2016 *IOP Conf. Ser.: Mater. Sci. and Engineering* **122** 012027. doi: 10.1088/1757-899X/155/1/012027.
[22] Karpov I, Ushakov A, Fedorov L and Lepeshev A 2014 *Technical Phys.* **84** 559. doi: 10.1134/S1063784214040148.
[23] Karpov I, Ushakov A, Lepeshev A and Fedorov L 2017 *Technical Phys.* **62** 1. doi: 10.1134/S106378421701011X.
[24] Saburov V, Cherepanov A, Zhurov M 1995 Plasma chemical synthesis of ultrafine powders and their application for the modification of metals and alloys. (Novosibirsk : RAS Siberian Publishing House) 344 p.
[25] Aksenov I, Brenn V, Padalka V, Khoroshikh V 1978 *Technical Phys* **49** 1165.
[26] Andrievsky R, Nuzhdin A 1986 Amorphous and ultra-fine powders and materials on their basis. Results of science and technology. Series: Powder metallurgy. (Moscow: Metallurgy) **2** 3.

IOP Conf. Series: Materials Science and Engineering **255** (2017) 012007 doi:10.1088/1757-899X/255/1/012007

The influence of magnetic field on the rate of cathode erosion at vacuum arc spraying

I V Karpov[1,2], A V Ushakov[1,2], A A Lepeshev[1,2], L Yu Fedorov[1,2], E A Dorozhkina[3], O N Karpova[3], A A Shaikhadinov[1,2], V G Demin[1,2], E P Bachurina[1,2], D V Lichargin[1], A K Abkaryan[2], G M Zeer[2], S M Zharkov[2,4]
[1]Federal Research Center "Krasnoyarsk Scientific Center" of the Russian Academy of Sciences, Krasnoyarsk, 660036, Russia
[2]Siberian Federal University, Krasnoyarsk, 660041, Russia
[3]Reshetnev Siberian State University of Science and Technology, Krasnoyarsk, 660037, Russia
[4]Kirensky Institute of Physics, Federal Research Center KSC SB RAS, Russia

E-mail: sfu-unesco@mail.ru

Annotation. The influence of the magnetic field in the cathode space on the synthesis of metal oxide nanopowders by vacuum-arc spraying was studied. It was found that, depending on the geometry of the magnetic field and the pressure of gaseous medium in plasma-chemical reactor, particular conditions which enhance the efficiency of CuO synthesis, appear.

One of the basic investigation tasks in the development of nanoparticles synthesis method [1-23] is to study the dependence of the average size and morphology of nanoparticles on process parameters. It is necessary to determine and optimize the physical variables that influence on the nucleation and growth of particles in the vapor phase in order to understand the synthesis process.

Within the presented work, we carried out the experimental study of nanoparticles synthesis processes in order to identify the influence of the magnetic field in the cathode space on the efficiency of the vacuum-arc spraying method.

Although vacuum arc of cathodic shape with cold cathode is one of the most efficient sources of metal ions in respect of power consumption, however, there are opportunities to further enhance the degree of ionization of the plasma flow through the ionization of vapor components. The principal way of increasing the degree of ionization of the plasma flow is to use a special configuration of electric and magnetic fields in the volume of the plasma generator.

In the end generators of metals plasma various types of solenoids are used to create an external magnetic field. Solenoids create a field B in the system, which possesses a cylindrical symmetry [24]. This field is intended to stabilize the cathode spot (CS) on the end face of the cathode. However, it can be used for additional ionization and acceleration of plasma flow in the case of a correct choice of its size and configuration. Vector $\vec{B}$ in this case can be resolved into two components: B_r - radial and B_z - axial. An external electric field E between the cathode and the special cylindrical anode or metal walls of the generator (playing the role of the anode) also possesses a cylindrical symmetry. In general, the field lines of the fields E and B intersect so tha7t the angle between $\vec{E}$ a $\vec{B}$ is a function of coordinates z and r. Thus, we have a situation of movement of the plasma flow in crossed inhomogeneous electric and magnetic fields.

In general, the analysis of the plasma flow in the end low-current accelerator can be carried out within the MHD approximation.

To obtain the accelerated metal plasma for coating, the system parameters are selected from the following relations:

$$R_e << L << R_i;\ \omega_i\tau_i << 1,\ \omega_e\tau_e > 1;\ N = 1\text{-}100\ \text{kW}$$

where N is a power of electric arc; L is a characteristic length of the acceleration zone.

In this case, the plasma acceleration occurs mainly by magnetic forces, plasma electrons are strongly magnetized and ions slightly change their trajectory while moving through the acceleration zone L. The direction of discharge current around the cathode coincides with the direction of the force lines of the magnetic field, which greatly facilitates the emission of electrons into the plasma. Near the anode, the electrons must move across the force lines $\vec{B}$, which decreases their outflow and creates conditions for further ionization of the vapor component.

Charge separation during the particles move into the azimuth direction causes an azimuth current. The physical origin of this current is determined by the presence of the azimuth components of the Hall voltage in the corresponding equations of the MHD approximation. Thus, the value of azimuth current exceeds the value of radial and axial currents.

An accelerating force, acting on the plasma in the axial direction, is defined by equations (1-3):

$$F = \int_{S_2} \rho V_z^2 dS = F_1 - F_2, \quad (1)$$

$$F = \int_V j_\varphi B_z dSdz - \frac{1}{2}\int_V j_\varphi B_z r dS, \quad (2)$$

$$F_2 = P_1 S_1 + \int_0^L PdS, \quad (3)$$

where S_1 and S_2 are inlet and outlet cross-sections of the plasma jet acceleration zone; V is a volume of acceleration zone; P_1 is a pressure at the inlet of the acceleration zone.

The force F_1 is specified by interaction of the discharge current with the external magnetic field B and has an opposite force applied to the solenoid and the cathode. The force F_2 is of gas-dynamic origin and has an opposite force applied to the inner walls of the accelerator.

Thus, at the area of a diverging magnetic field, the Hall plasma acceleration zone appeared. The Hall accelerating force (1) can be represented as:

$$F = \alpha\beta_e I_d B_e \quad (4)$$

where a *is* a geometric factor. A rate of plasma outflow V under the action of this force is defined by the relation:

$$V \cong F / \dot{m} - \beta_e B_r, \quad (5)$$

where $\dot{m}$ is a mass flow rate of plasma.

IOP Conf. Series: Materials Science and Engineering **255** (2017) 012007 doi:10.1088/1757-899X/255/1/012007

Interaction of Hall current with a magnetic field component leads to the focusing of plasma jet along the axis of the system under the influence of a force $F_r = j_\varphi B_z$. Wherein, a component of ion current j_i along the Z axis is changed slightly due to $\beta >> 1$. However, under certain circumstances, due to the interaction of electrons with neutral atoms and ions, the integral spin of the plasma flow can be observed.

In the end Hall accelerator there is the effect of autoseparation of microdrop phase on the periphery of the plasma jet, which is apparently due to a negative charge appearing on the electrically insulated microdroplets in plasma.

The influence of longitudinal magnetic field on the character and rate of cathode erosion was studied by means of the installation described in [6, 20]. Copper of M0 grade was used as a cathode for spraying. In order to begin plasma chemical reactions, the chamber was preliminary evacuated to a pressure of $1.33 \cdot 10^{-3}$ Pa and then it was filled by gas mixture composed of 10% O_2 + 90% Ar. The synthesis was carried out at a pressure of gas mixture of 60 Pa. The value of the magnetic field generated by the solenoid at the cathode surface was varied from 40 to 80 Oe. A pulse arc with a frequency of 1 kHz, a pulse duration of 250 μs and an amplitude of the current in a pulse of 2.3 kA were supported by the power supply system.

Fig. 1 shows the experimental dependence of the magnetic field on the rate of cathode erosion and the voltage in the discharge gap. Increasing an intensity of the magnetic field causes decreasing an erosion rate by 1.8 times, while increasing a voltage in the discharge gap. The magnetic field influences on the character of the cathode spot motion, moving it from the center to the edges of the working face of the cathode. [25] Wherein uniform erosion of the entire working surface of the cathode is provided, while in the case without a magnetic field cathode erodes mainly in the central part, which is explained by the localization of the spot in the given area of the cathode under the influence of its own magnetic field of the arc. Extension of the area of cathode spot migration, under the influence of the solenoid field, increases the time during which the spot traverses the entire eroding surface of the cathode. The specific thermal load on the cathode surface decreases, which leads to some decrease in the rate of the cathode erosion by reducing the amount of generated droplets.

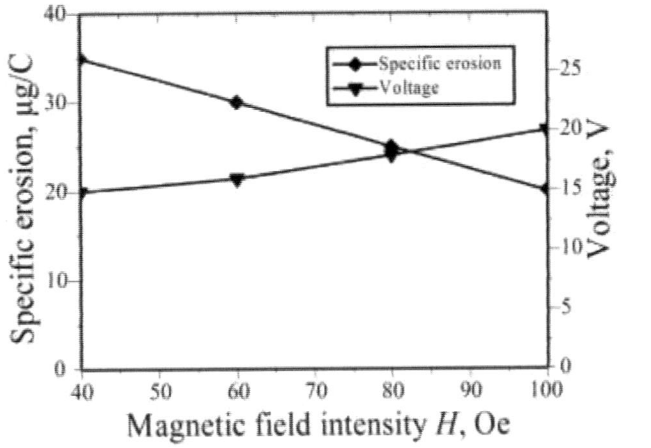

Figure 1. Dependence of the cathode specific erosion and the voltage in the discharge gap on the intensity of the longitudinal magnetic field

Under the condition of oxygen content in the volume, the expansion of spot migration area leads to increasing the time interval, during which at the sites of the cathode surface, which are not occupied by spot, copper oxide inclusions are formed, and hence the content of CuO on the cathode surface increases. When the cathode spot is situated on the copper oxide inclusions, the voltage drop at the discharge reduces. As a result, under the condition of sufficient density of CuO inclusions on the cathode surface, arcing occurs mainly on this material. Wherein, the magnetic field at the cathode does not influence the rate of the cathode erosion. The nature influence of the magnetic field in the volume on the rate of cathode erosion during the formation of copper oxide inclusions at its surface can be understood by assuming that active gas molecules enter into the reaction of CuO synthesis; and a basic elementary process, resulting in oxygen activation in these experimental conditions, is recharging copper ions on the gas molecules. This assumption is valid because of the presence in the vacuum arc plasma of a large number of multiply charged ions, exchange cross sections of which at low energies can be substantially higher than the gas-kinetic ones. Oxygen ions, appearing as a result of recharge process, have energies close to thermal and change their direction of motion under the presence of even weak electric fields in plasma. The paper [26] shows that in the presence of an external magnetic field, satisfying the condition: $\rho_i<\Lambda<<\rho_e$ (where ρ_i, ρ_e are ion and electron Larmor radius, respectively, Λ is a characteristic size of the system), electric field appears in vacuum arc plasma, and magnetic lines play role of equipotential lines of electric field [27].

Thus, while using the geometry of magnetic field, in the cathode space of plasma the electric field component appears, it directs toward the cathode working surface and accelerates oxygen ions into the given direction. At constant gas pressure in the system this effect causes increase of activated particles flow onto the working surface of the cathode, which causes an increase of efficiency of CuO synthesis, the formation of which determines decrease of the cathode erosion in the present case.

Acknowledgements. The work was performed with a support of the grant of the Russian Science Foundation (project no. 16-19-10054).

References

[1] Lepeshev A, Sordelet D, Rozhkova E and Ushakov A 2011 *J. of Cluster Sci.* **22** 289. doi: 10.1007/s10876-011-0378-2.

[2] Lepeshev A, Rozhkova E, Karpov I, Ushakov A and Fedorov L 2013 *Phys. of the Solid State* **55** 2531. doi: 10.1134/S1063783413120202.

[3] Ushakov A, Karpov I, Fedorov L and Lepeshev A 2014 *J of Friction and Wear* **35** 7. doi: 10.3103/S1068366614010103.

[4] Karpov I, Ushakov A, Fedorov L and Lepeshev A 2014 *Technical Phys.* **84** 559. doi: 10.1134/S1063784214040148.

[5] Ushakov A, Karpov I, Lepeshev A, Petrov M and Fedorov L 2014 *JETP Letters* **99** 99. doi: 10.1134/S002136401402009X.

[6] Ushakov A, Karpov I, Lepeshev A and Petrov M 2015 *J. Appl. Phys.* **118** 023907. http://dx.doi.org/10.1063/1.4926549.

[7] Ushakov A, Karpov I, Fedorov L, Lepeshev A, Shaikhadinov A, and Demin V 2015 *Theor. Foundations of Chemical Engineering*. **49** 743.

[8] Fedorov L, Karpov I, Ushakov A and Lepeshev A 2015 *Inorg. Mater.* **51** 25. doi: 10.1134/S0020168515010057.

[9] Lepeshev A, Bayukov O, Rozhkova E, Karpov I, Ushakov A and Fedorov L 2015 *Phys. of the Solid State* **57** 255. doi: 10.1134/S1063783415020249.

[10] Ushakov A, Karpov I, Lepeshev A, Petrov M and Fedorov L 2015 *Phys. of the Solid State* **57** 919. doi: 10.1134/S1063783415050303.
[11] Ushakov A, Karpov I and Lepeshev A 2015 *Phys. of the Solid State* **57** 2320. doi: 10.1134/S1063783415110359.
[12] Ushakov A, Karpov I, Lepeshev A, Fedorov L and Shaikhadinov A 2016 *Technical Phys.* **86**, 103. doi: 10.1134/S1063784216010230.
[13] Ushakov A, Karpov I and Lepeshev A 2016 *Technical Phys.* **86**, 260. doi: 10.1134/S1063784216020262.
[14] Lepeshev A, Karpov I, Ushakov A and Nagibin G 2016 *J. of Alloys and Compounds* **663** 631. doi:10.1016/j.jallcom.2015.12.168.
[15] Lepeshev A, Karpov I, Ushakov A, Fedorov L and Shaikhadinov A 2016 *Intern. J. of Nanoscience* **15** 1550027.
[16] Karpov I, Ushakov A, Lepeshev A and Zharkov S 2016 *Vacuum* **128** 123. doi: 10.1016/j.vacuum.2016.03.025.
[17] Ushakov A, Karpov I, Lepeshev A and Petrov M 2016 *Vacuum* **133** 25. doi: 10.1016/j.vacuum.2016.08.007.
[18] Lepeshev A, Rozhkova E, Karpov I, Ushakov A, Fedorov L, Dorozhkina E and Karpova O 2016 *IOP Conf. Series: Materials Science and Engineering* **155** 012014. doi: 10.1088/1757-899X/155/1/012014.
[19] Karpov I, Ushakov A, Lepeshev A, Dorozhkina E, Karpova O, Shaikhadinov A and Demin V 2016 *IOP Conf. Series: Materials Science and Engineering* **155** 012013. doi: 10.1088/1757-899X/155/1/012013.
[20] Karpov I, Ushakov A, Lepeshev A and Fedorov L 2017 *Technical Phys.* **62** 1. doi: 10.1134/S106378421701011X.
[21] Ushakov A, Karpov I and Lepeshev A 2017 *Journal of Superconductivity and Novel Magnetism* **30** 311. doi: 10.1007/s10948-016-3709-6.
[22] Rudenko K, Miakonkih A, Rogojin A, Bogdanov S, Sidorov V and Zelenkov P 2016 *IOP Conf. Ser.: Mater. Sci. and Engineering* **122** 012029. doi: 10.1088/1757-899X/122/1/012029.
[23] Bogdanov S, Lelekov E, Kovalev I, Zelenkov P and Lelekov A 2016 *IOP Conf. Ser.: Mater. Sci. and Engineering* **122** 012027. doi: 10.1088/1757-899X/155/1/012027.
[24] Grishin S, Leskov L, Kozlov N 1983 *Plasma accelerators*, 231 p.
[25] Aksenov I, Andreev A 1977 *Technical Physics Letters.* **3** 1272.
[26] Aksenov I, Belous V, Padalka V, Khoroshikh V 1978 *Plasma Physics.* **4** 758.
[27] Morozov A 1965 Focusing cold quasi-neutral beams in electromagnetic fields. Reports of the USSR Academy of Sciences. **163** 1363.

The investigation of the influence of oxygen concentration in the gas mixture on nanodispersed oxides synthesis

I V Karpov[1,2], A V Ushakov[1,2], A A Lepeshev[1,2], L Yu Fedorov[1,2], E A Dorozhkina[3], O N Karpova[3], A A Shaikhadinov[1,2], V G Demin[1,2], E P Bachurina[1,2], D V Lichargin[1], A K Abkaryan[2], G M Zeer[2], S M Zharkov[2,4]
[1]Federal Research Center "Krasnoyarsk Scientific Center" of the Russian Academy of Sciences, Krasnoyarsk, 660036, Russia
[2]Siberian Federal University, Krasnoyarsk, 660041, Russia
[3]Reshetnev Siberian State University of Science and Technology, Krasnoyarsk, 660037, Russia
[4]Kirensky Institute of Physics, Federal Research Center KSC SB RAS, Russia

E-mail: sfu-unesco@mail.ru

Annotation. Thermal effects in cathode space of low-pressure arc discharge, which contribute to the synthesis of nanopowders with an average particle size of less than 10 nm, are considered. One of the most important parameters, characterizing the cathode processes of vacuum arc, is a voltage drop across the discharge gap. All the gases demonstrate a decrease of a voltage drop across the discharge gap under the condition ofincreasing the gas pressure in the range of $p = 10^{-3} - 1$ Pa. Dissipation of energy of ions and electrons on gas molecules causes abrupt heating of the gas. Heat, coming from gas plasma, has a significant impact on evaporating material of a cathode. It can be assumed that a microdrop fraction, formed as a result of a cathode spraying in a liquid phase, will further evaporate in a superheated gas.

The papers [1, 2] describe a method for production of nanopowders (NP) of metals and their compounds with non-metals in the plasma of a pulsed low pressure acr discharge. Disperse powder composition is formed by means of two processes: dispersion-cooling of liquid metal and evaporation - condensation of steam phase. The studies [3-23] have shown that the obtained powder particles represent single crystals with a spherical shape. The average particle size was 10 nm.

The results clearly indicate a qualitative difference of powders, produced at different pressures of the gas mixture. The empirical formula are proposed for the dependence of a number-average diameter of the particle d_{cp}ona pressure p:

$$d_{cp} = 0.3\ 10^{-6}p^{-3},\ \text{m, at } p < 10 \text{ Pa};\ d_{cp}= 0.5\ 10^{-6}p^{-0,5},\ \text{m, at } p > 10 \text{ Pa}$$

While changing the pressure in the plasma-chemical reactor, the distribution function does not change, indicating that the nature of the synthesis ofelectric arc powders is predominantly thermal,and particles form by condensation from vapor phase.

One of the main factors, determining the conditions of the chemical reaction, is proportion of plasma-generating and reactive gases in the plasma-chemical rector. At constant values of the parameters of electrical vacuum arc (EVA) and the pressure of the gaseous medium, the amount of the formed oxide is determined by the oxygen content in the gas.

Let's consider the influence of this factor on the synthesis of copper oxide, taking into account the voltage drop across the discharge gap. Among the oxygen-contained components the most appropriate are air and oxygen of technical purity. The studies were carried out using the device described in [3] at a discharge current of 300 A, an intensity of the longitudinal magnetic field induced by the focusing coil on the surface of the cathode, of 80 Oe. Copper of electrolytic refining M0 was used as a cathode for spraying. The chamber was preliminary evacuated to a pressure of 1 mPa and then it was filled by gas mixtures based on argon, oxygen and air.

The dependence of the voltage drop *U*on the gas pressure for the copper cathode is shown in Fig. 1 and Fig. 2. It is evident for the gas mixtures of argon-oxygen and argon-air, that the increase of the pressure in the range of p = 1–10 Pa the decrease of the voltage drop across the discharge gap. Within increasing pressure of gas mixture higher than 10 Pa, a slight increasing*U*value is observed, which stabilizes at p = 15 Pa. Presence of oxygen and air in the system more than 5%, at $p>$ 15 Pa, causes a sharp, almost abrupt increasing U value. While further increase of pressure above 60–70 Pa the system demonstrates instabilities and frequent interruptions of the arc discharge. For gas mixtures based on oxygen the optimal gas mixture consists of 10% O_2 + 90% Ar, the limit pressure of stable arc discharge is 60 Pa. For gas mixtures based on air the optimal gas mixture consists of 20% air + 80% Ar, the limit pressure of stable arc discharge is 70 Pa.

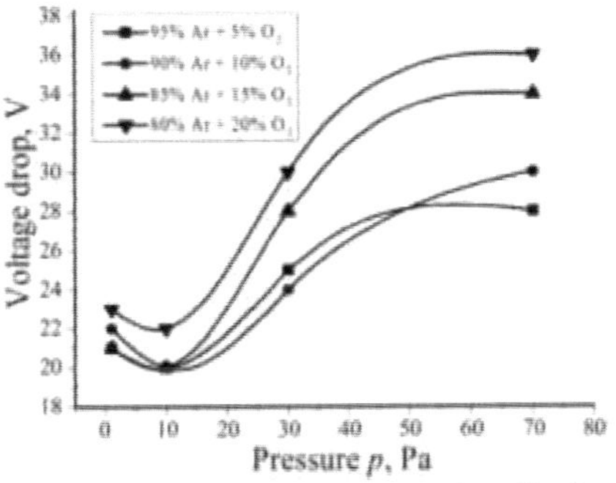

Figure 1. The dependence of the voltage drop across the discharge gap on the pressure of the gas mixture of argon and oxygen for copper cathode

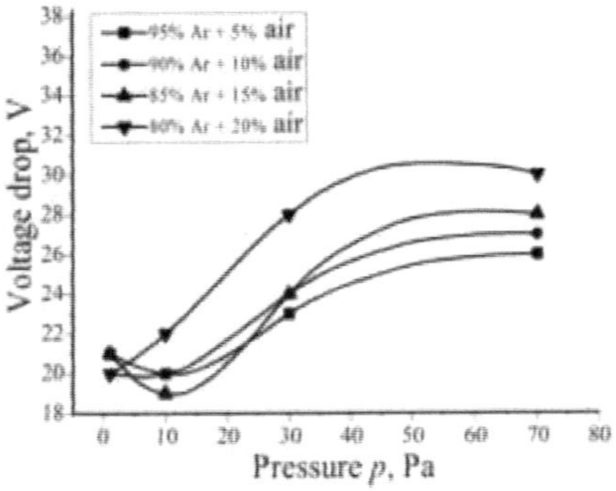

Figure 2. The dependence of the voltage drop across the discharge gap on the pressure of the gas mixture of argon and air for copper cathode

The voltage drop across the discharge gap consists of the cathode drop and the voltage drop on the positive pole of the arc. The change of each component with the increase of p value may have a different character and, therefore, different impact on the curves which image U values. Increase of U value under the pressure of the gas mixture $p > 10^{-1}$ Pa is caused by the processes in the arc column, which result in a decrease of conductivity of the plasma [24]: $\sigma = \frac{e^2 \cdot n_e}{m \cdot v_{eff}}$, where n_e, m – concentration and mass of the electron respectively; v_{eff}– effective frequency of collisions between electrons and atoms.

Wherein, decrease of U valueis mainly related with an increase of frequency of electron collisions with neutrals under the increase of gas pressure in the volume. In the system copper-oxygen and copper-air a sharp decrease in U value can also be associated with increase of average frequency of ion charge. These ions are generated during the formation of chemical compounds on the surface of the cathode by cathodespot of the arc. Dissipation of energy of ions and electrons on gas molecules causes abrupt heating of the gas. Moreover, the increased gas concentration let to assume a significant energy content of gas in the cathode space. The heat, coming from the gas plasma, significantly influences the evaporated cathode material. It can be assumed, that the microdrop fraction, formed as a result of cathode spraying in the liquid phase, will further evaporate in the superheated gas. As known for low-temperature plasma of atmospheric pressure, the nature and intensity of the heat and mass transfer between the particles of the dispersed phase and the plasma flow will be determined by energy content of the plasma flow, the physicochemical properties of the fine phase particles,type of plasma-forming gas and, finally, a character of interaction between the plasma flow andthefine-dispersed phase, which mostly depends on the Knudsen criterion (Kn).

Heat and mass transfer in the case of Kn<0.1 (high-pressure plasma) is most fully studied in the literature [25]. Wherein heat exchange is caused mainly by thermal conductivity and convection, and it can be described by the criterion equations, using the criteria of Nusselt, Reynolds, Prandtl, Lewis, etc. At the same time, when the character of plasma flow becomes molecular (Kn> 1), the convection processes are insignificant.Under these conditions, heat and mass transfer processes can be described by the principles of the kinetic theory. In this regard, to estimate the possibility of evaporation of microdrop fraction particles in the cathode plasma, a model based on the kinetic approach wasdeveloped. This model takes into account the fact that there is a flow of fast electrons in the cathode plasma, which are accelerated in the double electrostatic layer arising in the cathodic side of narrowing in a compressing magnetic field. The developed mathematical model includesthe equations of power balance of energy flows, which influence a small-sized solid body, placed into the plasma of low or medium pressure, and the equations describing non-stationary heat and mass transfer in a spherical body of small size, assuming the absence of temperature gradients within the body (body was considered as isothermal). The plasma parameters in the arc evaporator, necessary for simulation, were obtained experimentally [6]. Thus, it was found that under the argon pressure in the evaporator equal to ~ 34 Pa and an anode current of about 500 A, an electron temperature is equal to $T_e = 10^4$ K, and aconcentration of plasma is equal to 10^{21} m^{-3}.

The results show that the main flow of heat from the plasma to the particle is due to the energy brought by the ions. The part of the energy, brought by the electrons in the considered electronic temperature range, does not exceed 25%. Under the temperature of evaporation, the power from the powder particle is taken away entirely due to thermal radiation. The maximal temperature, to which a particle, introduced into the plasma, can be heated, is determined by the energy balance. In the above case, the copper particle temperature does not exceed 3000 K. Estimates of a time of complete evaporation of the copper particles $\tau_{нсп}$ with a diameter of 55 μm at $T_c = 10^4$ K gave the value of 10^{-5} s, which is much less than the time of the copper particles existence near the plasma column in the arc evaporation at its rate of its transfer across the flow not exceeding 100 m/s.

Thus, the estimations show that, under the condition of the sufficient energy content of the plasma flow, the process of cathode evaporation in low pressure arc discharge consists of two stages: firstly the cathode material is sprayed in the liquid phase in the cathode spot, and then is completely evaporated in the near-electrode gas-vapor mixture. Furthermore, due to mixing of metal vapors with a flow of ionized carrier gas, vapors are overheated, thus preventing premature condensation of vapors and causing dissociation of the already formed clusters.

Acknowledgements. The reported study was funded by Russian Foundation for Basic Research, project no 17-48-240806, Government of Krasnoyarsk Territory, Krasnoyarsk Region Science and Technology Support Fund to the Agreement №14.

References

[1] Karpov I, Ushakov A, Fedorov L and Lepeshev A 2014 *Technical Phys.* **84** 559. doi: 10.1134/S1063784214040148.
[2] Karpov I, Ushakov A, Lepeshev A and Fedorov L 2017 *Technical Phys.* **62** 1. doi: 10.1134/S106378421701011X.
[3] Lepeshev A, Sordelet D, Rozhkova E and Ushakov A 2011 *J. of Cluster Sci.* **22** 289. doi: 10.1007/s10876-011-0378-2.
[4] Lepeshev A, Rozhkova E, Karpov I, Ushakov A and Fedorov L 2013 *Phys. of the Solid State* **55** 2531. doi:10.1134/S1063783413120202.
[5] Ushakov A, Karpov I, Fedorov L and Lepeshev A 2014 *J of Friction and Wear* **35** 7. doi: 10.3103/S1068366614010103.
[6] Ushakov A, Karpov I, Lepeshev A, Petrov M and Fedorov L 2014 *JETP Letters* **99** 99. doi: 10.1134/S002136401402009X.
[7] Ushakov A, Karpov I, Lepeshev A and Petrov M 2015 *J. Appl. Phys.* **118** 023907. http://dx.doi.org/10.1063/1.4926549.
[8] Ushakov A, Karpov I, Fedorov L, Lepeshev A, Shaikhadinov A, and Demin V 2015 *Theor. Foundations of Chemical Engineering*.**49** 743.
[9] Fedorov L, Karpov I, Ushakov A and Lepeshev A 2015 *Inorg. Mater.***51** 25.doi: 10.1134/S0020168515010057.
[10] Lepeshev A, Bayukov O, Rozhkova E, Karpov I, Ushakov A and Fedorov L 2015 *Phys. of the Solid State***57** 255. doi: 10.1134/S1063783415020249.
[11] Ushakov A, Karpov I, Lepeshev A, Petrov M and Fedorov L 2015 *Phys. of the Solid State* **57** 919. doi: 10.1134/S1063783415050303.
[12] Ushakov A, Karpov I and Lepeshev A 2015 *Phys. of the Solid State* **57** 2320. doi: 10.1134/S1063783415110359.
[13] Ushakov A, Karpov I, Lepeshev A, Fedorov L and Shaikhadinov A 2016 *Technical Phys.* **86**, 103. doi: 10.1134/S1063784216010230.
[14] Ushakov A, Karpov I and Lepeshev A 2016 *Technical Phys.* **86**, 260. doi: 10.1134/S1063784216020262.
[15] Lepeshev A, Karpov I, Ushakov A and Nagibin G 2016*J. of Alloys and Compounds* **663** 631. doi:10.1016/j.jallcom.2015.12.168.
[16] Lepeshev A, Karpov I, Ushakov A, Fedorov L and Shaikhadinov A 2016 *Intern. J. of Nanoscience* **15** 1550027.

[17] Karpov I, Ushakov A, Lepeshev A and Zharkov S 2016 *Vacuum* **128** 123. doi: 10.1016/j.vacuum.2016.03.025.
[18] Ushakov A, Karpov I, Lepeshev A and Petrov M 2016 *Vacuum* **133** 25. doi: 10.1016/j.vacuum.2016.08.007.
[19] Lepeshev A, Rozhkova E, Karpov I, Ushakov A, Fedorov L, Dorozhkina E and Karpova O 2016 *IOP Conf. Series: MaterialsScienceand Engineering* **155** 012014. doi:10.1088/1757-899X/155/1/012014.
[20] Karpov I, Ushakov A, Lepeshev A, Dorozhkina E, Karpova O, Shaikhadinov A and Demin V 2016 *IOP Conf. Series: MaterialsScienceand Engineering* **155** 012013. doi:10.1088/1757-899X/155/1/012013.
[21] Ushakov A, Karpov I and Lepeshev A 2017 *Journal of Superconductivity and Novel Magnetism* **30** 311. doi:10.1007/s10948-016-3709-6.
[22] Rudenko K, Miakonkih A, Rogojin A, Bogdanov S, Sidorov V and Zelenkov P 2016 *IOP Conf. Ser.: Mater. Sci. and Engineering* **122** 012029. doi: 10.1088/1757-899X/122/1/012029.
[23] Bogdanov S, Lelekov E, Kovalev I, Zelenkov P and Lelekov A 2016 *IOP Conf. Ser.: Mater. Sci. and Engineering* **122** 012027. doi: 10.1088/1757-899X/155/1/012027.
[24] SpitzerL1957 Physics of Fully Ionized Gases. M.: Publishing House of Foreign Lit. 316 p.
[25] Kimblin S 1977 Experimental studies of plasma torches (Novosibirsk: Nauka) p. 226-253.

IOP Conf. Series: Materials Science and Engineering **255** (2017) 012009 doi:10.1088/1757-899X/255/1/012009

An adaptive algorithm for interference suppression in phased antenna arrays*

I N Kartsan[1]**, V N Tyapkin**[2]**, D D Dmitriev**[3]**, A E Goncharov**[4]**, I V Kovalev**[5]
[1]Associate professor, Reshetnev Siberian State University of Science and Technology, Krasnoyarsk, Russia
[2]Associate professor, Siberian Federal University, Krasnoyarsk, Russia
[3]Associate professor, Siberian Federal University, Krasnoyarsk, Russia
[4]Associate professor, Reshetnev Siberian State University of Science and Technology, Krasnoyarsk, Russia
[5]Professor, Reshetnev Siberian State University of Science and Technology, Krasnoyarsk, Russia

E-mail: kartsan2003@mail.ru

Abstract. This paper describes an algorithm designed for adapting a phased antenna array to an interference environment. Based on this algorithm, the authors have proposed a design of an adaptive beamformer.

1. Introduction

Phased antenna arrays are becoming an essential part of many modern electronic devices. Despite their complicated design and relatively bulky appearance, phased antenna arrays have one major advantage over simple antennas – it is possible to control the shape of their radiation pattern.

Typically, radiation patterns are controlled in two situations: the forming of minimums in direction to the source of interference and maximums in direction of the source or sources of the desired signal; this includes shifting signals. If a phased antenna array is used in the design of command and measurement systems of ground control facilities for spacecraft, both of these situations are considered.

Phased antenna arrays significantly improve such technical and economical performance factors of command and measurement systems of ground control facilities for spacecraft as:

- increasing the number of spacecraft which are tracked simultaneously from the ground;
- providing protection from all types of electronic interference using adaptive spatial filtering algorithms;
- improving the operation efficiency of data exchange from the satellite and the performance of the technological operations; reducing the time interval for performing a technological control cycle;
- improving the reliability of the measurement systems of ground control facilities by excluding mechanical elements from the actuator of the antenna system.

It is considered that among the more perspective methods for suppressing interference are

* The work was supported by the Ministry of Education and Science of the Russian Federation (agreement from 03.10.2016 № 14.577.21.0220, unique ID project RFMEFI57716X0220)

adaptive spatial filtering algorithms [1]; such algorithms are used in phased antenna arrays. Theoretical issues behind the design and application of phased antenna arrays have been widely discussed in both Russian and international publications. However the application of phased antenna arrays in measurement systems of ground control facilities for spacecraft has a number of features, which require further investigation. These features include a low level of the desired signal, a requirement for receiving desires signals from different directions, and the high speed of the satellite.

In [2] it had been demonstrated that when the weighted vector of the phased antenna array is calculated, the choice of the criterion for determining the efficiency does not matter. It is more important to select an appropriate control algorithm for adjusting the radiation pattern of the phased antenna array; this algorithm directly influences the speed of the transition process and the complexity of the whole system. In [3, 4] an adaptive phased antenna array with initial uniform amplitude distribution has been described; the efficiency of its adaptive algorithm has been determined using a basis in the form of a partial exponential Fourier row in comparison to an optimal algorithm. In this paper we shall perform a synthesis of an adaptive algorithm for an 8-element circular phased antenna array.

2. A synthesis of an adaptive algorithm for an 8-element circular phased antenna array

The received signal generally follows the vector function, which considers its specification in time and space. If a signal is extracted during interference, the differences between them are used. From a mathematical point of view, it is convenient to consider these differences by means of the dependence of signals from interference and other spatial or time-frequency factors. Thus, the input oscillation shall be

$$y(t) = x(t, \alpha, \beta) + n(t, \nu), \tag{1}$$

where $x(t, \alpha, \beta)$ is the vector of the input desired signal, having parameters α and β; α is the vector of the informative signal parameters (phase, delay, Doppler shift, etc.); β is the vector of noninformative parameters, produced by signal fluctuations; $n(t, \nu)$ is the vector of interference-induced oscillation; ν is the vector of the interference parameters.

In terms of the interference, let us suppose that it is a vectorial accidental stationary process. The normal law of distributing interference-induced oscillation is physically justified in the majority of cases since the interference signal becomes normal in the relatively narrowband frequency paths of the receivers.

The phase of the input signal in each antenna element of the array is determined by the coordinates of this element, the direction of the source of the signals and interference, and the angle of the beam.

In order to determine the vector of the amplitude and phase distribution of the signal, it is necessary to know the direction, from which the signal arrives, the coordinates of the antenna elements and their radiation pattern.

The direction from which the signal arrives can be described by a single vector from a selected antenna element or phased center of the array to the signal source [5]

$$L(\psi,\gamma)= (\cos(\psi)\cos(\psi); \sin(\psi)\cos(\gamma); \sin(\psi))^{\mathrm{T}}, \tag{2}$$

where ψ is the azimuth to the signal source (navigation satellite); γ is the angular altitude to the signal source.

The structure of the phased antenna array is dictated by matrix A in which the k-th column contains orthogonal coordinates, y, z i-th antenna element with respect to the phase center of the array, which matches the circumference center of radius R; the antenna elements are situated on this circumference.

$$\mathbf{A} = R \cdot \begin{vmatrix} 0 & \frac{\sqrt{2}}{2} & 1 & \frac{\sqrt{2}}{2} & 0 & -\frac{\sqrt{2}}{2} & -1 & -\frac{\sqrt{2}}{2} \\ 1 & \frac{\sqrt{2}}{2} & 0 & -\frac{\sqrt{2}}{2} & -1 & -\frac{\sqrt{2}}{2} & 0 & \frac{\sqrt{2}}{2} \\ 0 & 0 & 0 & 0 & 0 & 0 & 0 & 0 \end{vmatrix}, \qquad (3)$$

Let us suppose that the elements of the phased antenna array are omnidirectional and their positioning is shown in Fig. 1.

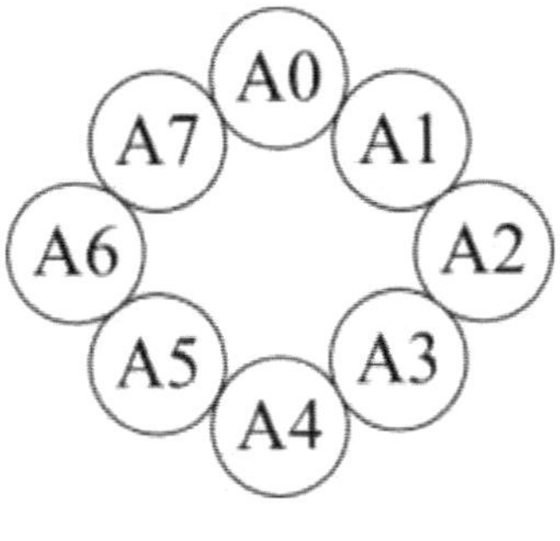

Figure 1. The position of antenna elements on the array.

The vector from the phase center of the array (its coordinates, in our case, are $(0\ 0\ 0)^{T}$) to the k-th element can be obtained as a difference between the coordinates of the circumference center and the array elements. In this case, this is, simply, the k-th column of matrix **A**, i.e. $\mathbf{A}^{<k>}$

Thus, the amplitude and phase distribution vector to the array elements, setting the direction of the beam to the i-th signal source, can be determined by

$$\mathbf{H}(\psi,\gamma) = \left(F_1(\psi,\gamma)\cdot\exp\left\{-2\pi j\cdot(R\mathbf{A}^{<1>}, L(\psi,\gamma))\right\}\ldots F_8(\psi,\gamma)\cdot\exp\left\{-2\pi j\cdot(R\mathbf{A}^{<8>}, L(\psi,\gamma))\right\}\right), \qquad (4)$$

where $F_1(\psi,\gamma)$ is the amplitude and phase radiation pattern of the k-th array element; $(R\mathbf{A}^{<1>}, L(\psi,\gamma))$ is the scalar (dot) product.

The total output signal from all the antenna elements of the phased antenna array can be determined by

$$\dot{\mathbf{Y}}_{\Sigma}(t) = \dot{\mathbf{Y}}^{T}\left[\dot{\mathbf{\Phi}}^{-1}\dot{\mathbf{X}}(\alpha)\right] = \dot{\mathbf{Y}}^{T}(t)\dot{\mathbf{R}}(\alpha), \qquad (5)$$

where $\dot{\mathbf{Y}}_\Sigma(t)$ is the output signal of the adaptive antenna array; $\dot{\mathbf{Y}}^T$ is the vector of the input signal of the array; $\dot{\mathbf{\Phi}}^{-1}$ is the inverse correlation of the interference matrix; $\dot{\mathbf{R}}(\alpha)$ is the vector of weighted coefficients.

The algorithm of spatial processing (5) is typically modified. This is done considering the requirements of a more convenient arrangement of the spatial processing unit. The modification of the algorithm is reduced to the multiplication of column vector of the received $\dot{\mathbf{Y}}(t)$ and expected $\dot{\mathbf{X}}(\alpha)$ oscillations by a transformation matrix **A** [6]. The transformation of the column vector can be determined using

$$\dot{\mathbf{Y}}_{tr}(t) = \dot{\mathbf{A}}\dot{\mathbf{Y}}(t), \quad \dot{\mathbf{X}}_{tr}(\alpha) = \dot{\mathbf{A}}\dot{\mathbf{X}}(\alpha). \tag{6}$$

Accordingly, the correlation matrix $\dot{\mathbf{\Phi}}$ and its opposite $\dot{\mathbf{\Phi}}^{-1}$ are modified, as is the weighted column vector $\dot{\mathbf{R}}(\alpha)$. Their modified values with considerations for (6) shall be equal to

$$\begin{gathered} \dot{\mathbf{\Phi}}_{tr} = M_{II}\left[\frac{\dot{\mathbf{Y}}_{tr}(t)\dot{\mathbf{Y}}_{tr}^{*T}(t)}{2}\right] = \dot{\mathbf{A}}\dot{\mathbf{\Phi}}\dot{\mathbf{A}}^{*T}, \\ \dot{\mathbf{\Phi}}_{tr}^{-1} = (\dot{\mathbf{A}}\dot{\mathbf{\Phi}}\dot{\mathbf{A}}^{*T})^{-1}, \quad \dot{\mathbf{R}}_{tr}(\alpha) = \dot{\mathbf{\Phi}}_{tr}^{-1}\dot{\mathbf{X}}_{tr}(\alpha). \end{gathered} \tag{7}$$

Furthermore, the structure of modifying the spatial processing algorithm is identical to the modified algorithm

$$\dot{\mathbf{Y}}_{\Sigma tr}(t) = \dot{\mathbf{Y}}_{tr}^T(t)\dot{\mathbf{R}}_{tr}(\alpha) = \dot{\mathbf{Y}}_{tr}(t). \tag{8}$$

When comparing algorithms (5) and (8) it is easy to become assured in their total equivalence if the transformation matrix **A** has an opposite $\dot{\mathbf{A}}^{-1}$. Actually, if in (8) we replace values $\dot{\mathbf{Y}}_{tr}(t)$ and $\dot{\mathbf{R}}_{tr}(\alpha)$ from (6) and (7), we shall get

$$\dot{\mathbf{Y}}_{tr}(t) = \dot{\mathbf{A}}^T(\dot{\mathbf{A}}^T)^{-1}(\dot{\mathbf{\Phi}}^{-1})^*(\dot{\mathbf{A}}^*)^{-1}\mathbf{A}^*\dot{\mathbf{X}}(\alpha) = \dot{\mathbf{Y}}(t)\dot{\mathbf{R}}^*(\alpha) = \dot{\mathbf{Y}}_\Sigma(t). \tag{9}$$

Matrix $\dot{\mathbf{A}}$ is used for building schemes with a main and (M – 1) compensating receiving channels, having near-omnidirectional features. The choice of the transformation matrix is determined by the decoupling of the primary and compensating channels via the desired signal. Here is one of the variants of such a matrix:

IOP Conf. Series: Materials Science and Engineering 255 (2017) 012009 doi:10.1088/1757-899X/255/1/012009

$$\dot{\mathbf{A}} = \begin{Vmatrix} \dot{\mathbf{X}}_1^*(\alpha) & \dot{\mathbf{X}}_2^*(\alpha) & \dot{\mathbf{X}}_3^*(\alpha) & \cdots & \dot{\mathbf{X}}_{m-1}^*(\alpha) & \dot{\mathbf{X}}_m^*(\alpha) \\ -\dot{\mathbf{X}}_1^*(\alpha) & \dot{\mathbf{X}}_2^*(\alpha) & 0 & \cdots & 0 & 0 \\ 0 & -\dot{\mathbf{X}}_2^*(\alpha) & \dot{\mathbf{X}}_3^*(\alpha) & \cdots & 0 & 0 \\ \cdots & \cdots & \cdots & \cdots & \cdots & \cdots \\ 0 & 0 & 0 & \cdots & -\dot{\mathbf{X}}_{m-1}^*(\alpha) & \dot{\mathbf{X}}_m^*(\alpha) \end{Vmatrix}. \quad (10)$$

The modified column vector $\dot{\mathbf{Y}}_{tr}(t) = \left\| \dot{\mathbf{Y}}_{tri}(t) \right\|$ of the received oscillation $\dot{\mathbf{Y}}(t)$ includes the strength of the primary channel

$$\dot{\mathbf{Y}}_{tr1}(t) = \sum_{i=1}^{M} \dot{\mathbf{Y}}_i(t) \dot{\mathbf{X}}_i^*(\alpha). \quad (11)$$

and the strength of the compensating channels

$$\dot{\mathbf{Y}}_{trki}(t) = -\dot{\mathbf{Y}}_i(t) \dot{\mathbf{X}}_i^*(\alpha) + \dot{\mathbf{Y}}_{i+1}(t) \dot{\mathbf{X}}_{i+1}^*(\alpha). \quad (12)$$

The anticipated oscillation $\dot{\mathbf{X}}(\alpha)$ are modified to

$$\dot{\mathbf{X}}_{tr}(\alpha) = \dot{\mathbf{A}} \dot{\mathbf{X}}(\alpha) = \left\| M \quad 0 \quad 0 \quad \cdots \quad 0 \right\|^T. \quad (13)$$

where $M = \sum_{i=1}^{M} \dot{\mathbf{X}}_i^*(\alpha) \dot{\mathbf{X}}_i(\alpha)$.

The zero value of the elements of the column vector $\dot{\mathbf{X}}_{tr}(\alpha)$, starting from the second, confirm a decoupling of the primary and compensating channels via the desired signal. The latter is an instrument for forming compensating channels using difference schemes. Due to this, a null is formed in the combined characteristics of the directivity of the compensating channels. The angular position of the null corresponds with the anticipated input of the desired signal α.

The formula for modifying the weighted vector column of spatial processing $\dot{\mathbf{R}}_{tr}(\alpha)$ can be found using the matrix equation

$$\dot{\mathbf{\Phi}}_{tr} \dot{\mathbf{R}}_{tr}(\alpha) = \dot{\mathbf{X}}_{tr}(\alpha). \quad (14)$$

It is conveniently solved by writing $\dot{\mathbf{\Phi}}_{tr}$ and the vector column $\dot{\mathbf{R}}_{tr}(\alpha)$ as blocks

$$\dot{\mathbf{\Phi}}_{tr} = \begin{Vmatrix} \dot{\mathbf{\Phi}}_{tr11} & \dot{\mathbf{\Phi}}_{tr}^{*T} \\ \dot{\mathbf{\Phi}}_{tr1} & \dot{\mathbf{\Phi}}_{tr\kappa} \end{Vmatrix} \qquad \dot{\mathbf{R}}_{tr}(\alpha) = \left\| \dot{\mathbf{R}}_1 \quad \dot{\mathbf{R}}_2 \right\|^T, \quad (15)$$

where $\mathbf{\Phi}_{tr11}$ is the total dispersion of interference and noise at the output of the primary receiving channel

$$\dot{\mathbf{\Phi}}_{tr11} = M_{\Pi}\left[\frac{\dot{\mathbf{Y}}_{tr1}(t)\dot{\mathbf{Y}}^{*}_{tr1}(t)}{2}\right]; \tag{16}$$

$\dot{\mathbf{\Phi}}_{tr1}$ is the vector column of bilateral correlation moments of the interference voltage in the compensating and primary channels

$$\dot{\mathbf{\Phi}}_{tr1} = M_{\Pi}\left[\frac{\dot{\mathbf{Y}}_{trk}(t)\dot{\mathbf{Y}}^{*}_{tr1}(t)}{2}\right], \quad k = \overline{2M}; \tag{17}$$

$\dot{\mathbf{\Phi}}_{tr\kappa}$ is the correlation matrix of the interference of the compensating channels

$$\dot{\mathbf{\Phi}}_{trk} = M_{\Pi}\left[\frac{\dot{\mathbf{Y}}_{trk}(t)\dot{\mathbf{Y}}^{*T}_{trk}(t)}{2}\right]; \tag{18}$$

$\dot{\mathbf{R}}_1$ and $\dot{\mathbf{R}}_2$ are elements of the weighted vector of the primary channel and the column vector of the weighted coefficients of the compensating channels.

Substituting (15–18) in the matrix equation (14), we deduce a system of two equations

$$\begin{cases} \dot{\mathbf{\Phi}}_{tr11}\dot{\mathbf{R}}_1 + \dot{\mathbf{\Phi}}^{*T}_{tr1}\dot{\mathbf{R}}_2 = M \\ \dot{\mathbf{\Phi}}_{tr1}\dot{\mathbf{R}}_1 + \dot{\mathbf{\Phi}}_{trk}\dot{\mathbf{R}}_2 = 0 \end{cases}, \tag{19}$$

Solving this system in relation to $\dot{\mathbf{R}}_1$ and $\dot{\mathbf{R}}_2$, we shall determine

$$\begin{cases} \dot{\mathbf{R}}_1 = \dfrac{M}{(\dot{\mathbf{\Phi}}_{tr11} - \dot{\mathbf{\Phi}}^{*T}_{tr1}\dot{\mathbf{\Phi}}^{-1}_{trk}\dot{\mathbf{\Phi}}_{tr1})}, \\ \dot{\mathbf{R}}_2 = -\dot{\mathbf{\Phi}}^{-1}_{trk}\dot{\mathbf{\Phi}}_{tr1}\dot{\mathbf{R}}_1 \end{cases} \tag{20}$$

It can be seen from (20) that the first element of the weighted vector is real value and serves as a normalization factor. This factor may be set to one without the loss of performance in detection, passing on to the normalized weighted vector $\dfrac{\dot{\mathbf{R}}_{tr}(\alpha)}{\dot{\mathbf{R}}_1}$. According to (20) the latter shall be equal to

$$\dot{\mathbf{R}}_{\Delta} = \frac{\dot{\mathbf{R}}_2}{\mathbf{R}_1} = -\dot{\mathbf{\Phi}}_{trk}^{-1}\mathbf{\Phi}_{tr1}, \tag{21}$$

The output voltage of the processing unit, having a designated primary channel, is determined by

$$\dot{\mathbf{Y}}_{\Sigma} = \dot{\mathbf{Y}}_{tr1}(t) + \dot{\mathbf{Y}}_{trk}^{*T}(t)\dot{\mathbf{R}}_{\Delta}^{*}, \tag{22}$$

The interference-compensating unit for interference received by the side lobes of a radiation pattern is built in accordance with the synthesized algorithm (Fig. 2).

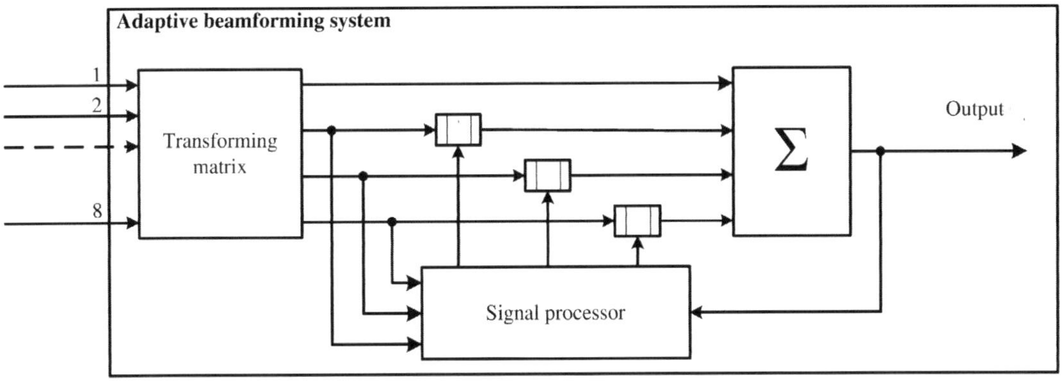

Figure 2. An adaptive beamforming system of a phased antenna array.

The analog section of the adaptive phased antenna array consists of 8 antenna modules with 8 connected radio signal paths. The signals from the radio signal paths are received by the analog-to-digital converter, after which they are transmitted to the digital unit, which consists of a beamforming arrangement and a digital signal processing channel. The beamforming arrangement is a weighted summer of signals, transmitted from 8 analog paths. The weighted coefficients are calculated by a signal processor in accordance with the desired radiation pattern. Particularly, in a high precision measuring mode, the weighted coefficients are selected so as to be capable of compensating the difference between the pass of the desired signal between antennas, i.e. the desired signals, transmitted to the summer, have one phase. This mode of the adaptive phased antenna array has been inquired in [7].

3. Conclusion

In conclusion it should be said that we have synthesized an interference-suppressing algorithm for an 8-element phased antenna array; this algorithm enables us to implement a close to optimal output effect of the adaptive unit. The formation of maximums in the radiation pattern in the direction of the signal source increases the signal-to-noise ratio, which, in its turn, improves interference immunity; the general increase in sensitivity aids receiving signals from satellites in difficult conditions.

References

[1] Radioelektronnie sistemy: osnovy postroeniia i teoriia [Radio-electronic systems: the fundamentals of design and theory (in Russ.)] 2007 Ia.D. Shirman (ed.) Moscow: Radiotekhnika.

[2] Tyapkin V N, Dmitriev D D, Fateev Y I, Kremez N S 2016 The synthesis algorithm for spatial filtering to maintain a constant level of the useful signal, *Journal of Siberian Federal University. Mathematics & Physics* **9(2)** pp 258–268.

[3] Kartsan I N, Tyapkin V N, Dmitriev D D, Goncharov A E, Zelenkov P V, Kovalev I V 2016 Synthesis of an algorithm for interference immunity *IOP Conference Series: Materials Science and Engineering* **155**

[4] Tyapkin V N, Kartsan I N, Dmitriev D D, Goncharov A E 2015 Spatial filtering algorithms in adaptive multi-beam hybrid reflector antennas, *International Siberian Conference on Control and Communications (SIBCON) – proceedings*

[5] GLONASS. Printsipy postroeniia i funktsionirovaniia [GLONASS: arrangement and functioning principles (in Russ.)] 2010, A I Perov, V N Kharisov (eds.), Moscow: Radiotekhnika

[6] Almazov V B, Manzhos V N 1985 Poluchenie i obrabotka radiolokatsionnoi informatsii [Receiving and processing radar data (in Russ.)] Kharkov

[7] Sharfunova T G, Tyapkin V N, Dmitriev D D 2013 Tochnost' izmereniia navagatsionnykh priborov v navigatsionnoi apparature potrebitelei sputnikovoi radionavigatsionnoi sistemy "GLONASS", osnashchennoi antennoi reshetkoi [The accuracy of navigation parameters measurements in GLONASS systems with an antenna array] *Radiotekhnika* **6** pp 22–25

IOP Conf. Series: Materials Science and Engineering 255 (2017) 012010 doi:10.1088/1757-899X/255/1/012010

Electromagnetic compatibility of ground system of near navigation, based on the use of GNSS and pseudolites

I N Kartsan[1], D D Dmitriev[2], A V Sokolovskii[3], V N Ratushnyak[4], A B Gladyshev[5], I V Kovalev[6]

[1]Associate professor, Reshetnev Siberian State University of Science and Technology, Krasnoyarsk, Russia
[2]Associate professor, Siberian Federal University, Krasnoyarsk, Russia
[3]Associate professor, Siberian Federal University, Krasnoyarsk, Russia
[4]Associate professor, Siberian Federal University, Krasnoyarsk, Russia
[5]Associate professor, Siberian Federal University, Krasnoyarsk, Russia
[6]Professor, Reshetnev Siberian State University of Science and Technology, Krasnoyarsk, Russia

E-mail: kartsan2003@mail.ru

Abstract. Article describes the results of modeling the allowable power pseudolites in the ground system of near navigation, working together with GLONASS. Was justified pulsed operation of pseudolites and using weighting windows to improve electromagnetic compatibility of pseudolites signals with GNSS signals.

1. Introduction

At present days, more and more attention is paid to radio communications or navigation systems. The most widely used is a global navigation satellite systems (GNSS). They are characterized by high availability and sufficient for ordinary consumers accuracy. However, the possibilities offered by GNSS is insufficient for activities requiring continuous high-precision navigation support. This is to ensure aviation safety in airport areas; ensuring flights of unmanned aircraft; navigation in port areas, the narrowness of the rivers; provision of geodetic, cartographic and special works. The reason for this is the low noise immunity of GNSS, and their lack of accuracy in the mountains or urban areas.

One of the ways to resolve this contradiction is augmentation of GNSS signals by ground-based radio navigation system near range navigation, which allows you to create a navigation field in a given region of space with the help of ground-based pseudolites [1, 2]. High precision measuring consumer navigation parameters in a radio navigation system provides high precision geodetic reference and mutual synchronization of pseudolites, which are stationary emitters navigation signals similar to the signals of navigation satellites (NS) GLONASS.

This short-range navigation architecture is significantly different from the local, regional or wideband differential GNSS subsystems. If the use of the latter is possible only with a stable receiving GLONASS signals, using a navigation system based on pseudolites can ensure the navigation even in the absence of signals from the NS. In addition, such system allows to ensure the required accuracy of measurement of coordinates and angles of spatial orientation of the consumer at the design stage [3].

One of the main problems faced in the development of such systems - is the problem of electromagnetic compatibility of pseudolites signals and navigation satellites. [4] In the operating frequency channel, it is necessary to implement them additional isolation. This can be done in three ways:

- add a small shift in operating frequency pseudolites within the working channel frequency;
- add artificial offset pseudorandom sequence ranging code pseudolites to ensure minimal cross-correlation between the signal pseudolite and GLONASS signal on the same frequency [5];
- use pulsed pseudolites mode. In this case the receiver must be made pseudolites blanking pulse in the channel receiving the signal on one frequency GLONASS [6].

To achieve the best cross-correlation may use a combination of these methods.

Together with EMC problems that must be solved in the local navigation augmentation systems based on pseudolites, also necessary to solve a "near-far" problem. This problem is that when an over-power signal pseudolites 20 dB and it creates more interference for receiving GLONASS signals. The same effect occurs when working with several pseudolites user approaching one of them.

2. Allowable power pseudolites calculation

Consider a pair of pseudolite – GLONASS satellite operating at the same frequency and define the parameters of the signals, at which it is possible to avoid the "near-far" effect. Navigation satellites signals received by the antenna apparatus of the consumer, have an average power:

$$P_{GNSS} = P_G - L_F - L_{AT} - L_{POL} - L_0 \approx -161\ldots-163 \text{ dBW}, \quad (1)$$

where P_G – effective transmitter power (28...30 dBW); L_F – free space loss $20\lg(\lambda/4\pi R) \approx 182.2...184.2$ dB for GLONASS signals); L_{AT} – atmospheric losses (≈ 2 dB); L_{POL} – polarization losses (≈ 1 dB); L_0 – other losses ($\approx 3...4$ dB).

Given that pseudolites signal loss associated with propagation in the atmosphere, and the polarization losses are negligible, the conditions under which the "near-far" effect can be avoided, is described by the inequality:

$$|P_{PL} - P_{GNSS}| \approx |PG_{TX,PL} - PG_{TX,GNSS} + 20\lg(R_{GNSS}/R_{PL}) - 3 \text{ dB}| < 20 \text{ dB}, \quad (2)$$

where P_{PL} – pseudolite transmitter power; P_{GNSS} – GNSS satellite transmitter power; $PG_{TX,PL}$ – pseudolite effective transmitter power; $PG_{TX,GNSS}$ – GNSS satellite effective transmitter power; R_{GNSS} – distance from the satellite to the GNSS consumer receiver equipment; R_{PL} – distance from the pseudolite to the GNSS consumer receiver equipment.

Thus we arrive, in order to eliminate the "near-far" effect in the entire range of possible power signals of the GLONASS space navigation devices must be observed the following inequality:

$$(11\ldots13) - PG_{TX,PL} < 10\lg(R_{GNSS}/R_{PL}) < (51\ldots53) - PG_{TX,PL}. \quad (3)$$

The range of power pseudolites signals by 2 dB is due to some scatter the power of the transmitted signal space navigation unit and the change in the value of its attenuation at different elevations. The results of modeling the allowable power pseudolite transmitter radiation range are shown in Fig. 1.

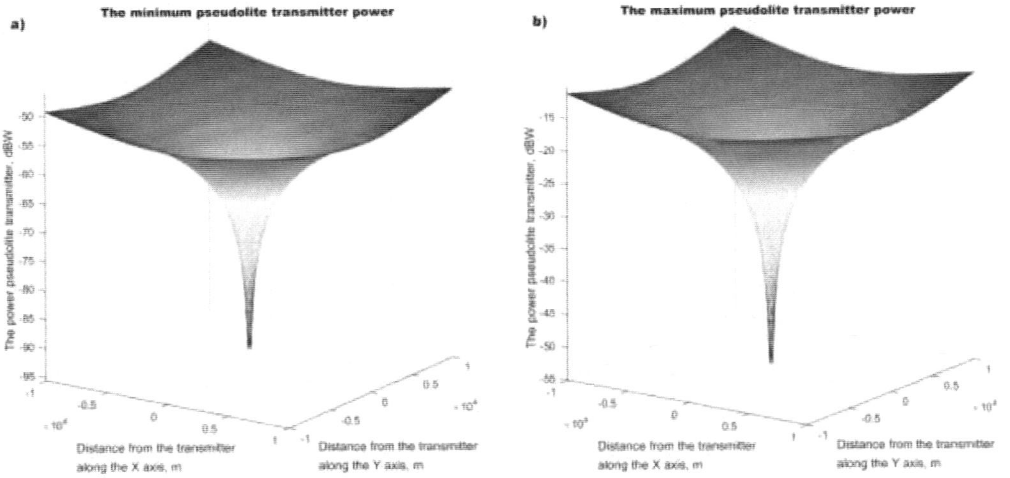

Figure 1. Dependence pseudolite allowable power of the transmitter of the distance in the plane of the Earth's surface: *a* – minimum power value; *b* – maximum power value.

You can see a big difference gradient power depending on the distance from the transmitter pseudolite that limits the working range of the signals pseudolite. To improve the performance of the consumer equipment pseudolites signals are usually transmitted in the pulse mode. To assess the degree of influence of pulse mode pseudolite transmission signal on the GLONASS signal, consider the correlation operation on the T accumulation interval in the *i*-th channel.

$$\begin{aligned} I_{i,PL}(t) &= \int_0^T h_{i,DC}(t) h_{PL}(t-\tau)\cos\left(\Delta\omega_{i,DC}t + \Delta\varphi_{i,DC}\right)dt, \\ Q_{i,PL}(t) &= \int_0^T h_{i,DC}(t) h_{PL}(t-\tau)\sin\left(\Delta\omega_{i,DC}t + \Delta\varphi_{i,DC}\right)dt, \end{aligned} \qquad (4)$$

where $h_{i,DC}$ – a pseudo-random sequence ranging code of *i*-th channel; h_{PL} – code sequence of the received signal; $\Delta\omega_{i,DC}$ – the difference between the local oscillator frequency and the frequency of the received signal *i*-th channel; $\Delta\varphi_{i,DC}$ – the phase difference between the local oscillator and the received signal phase *i*-th channel.

Let pseudolite signal is a weighted by Henning window (with default parameter $\alpha = 1$) normalized delta pulse with τ_p duration, the signal at the output of the correlator will be presented in (5).

For the analysis we assume that the received signal pseudolite and GLONASS signals are completely correlated with the code sequence of the receiver, but also do not have Doppler frequency shift.

IOP Conf. Series: Materials Science and Engineering **255** (2017) 012010 doi:10.1088/1757-899X/255/1/012010

$$I_{i,PL}(t)=\int_0^{\tau_p} h_{i,DC}(t)h_{PL}(t-\tau)\sin\left(\frac{\pi t}{\tau_p}\right)\cos\left(\Delta\omega_{i,DC}t+\Delta\varphi_{i,DC}\right)dt,$$
$$Q_{i,PL}(t)=\int_0^{\tau_p} h_{i,DC}(t)h_{PL}(t-\tau)\sin\left(\frac{\pi t}{\tau_p}\right)\sin\left(\Delta\omega_{i,DC}t+\Delta\varphi_{i,DC}\right)dt. \quad (5)$$

Then, according to (5) a pulse signal pseudolite and GLONASS correlated signals will have a real component only is equal to

$$I_{i,PL}(t)=\frac{\tau_p}{\pi}\sum_{k=1}^{\tau_p/\tau_e} h_{ik,DC}(t)h_{k,PL}(t-\tau)\left[\cos\left(\frac{(k-1)\pi\tau_e}{\tau_p}\right)-\cos\left(\frac{k\pi\tau_e}{\tau_p}\right)\right],$$
$$I_{i,GNSS}(t)=\sum_{k=1}^{T/\tau_e} h_{ik,DC}(t)h_{k,PL}(t-\tau), \quad (6)$$

where τ_p – pseudolite pulse duration; τ_e – duration of the chip code sequence.

System health condition (2) in view of (6) is equal to

$$\left|20\lg\frac{I_{i,PL}(t)}{I_{i,GNSS}(t)}+20\lg\frac{R_{GNSS}}{R_{PL}}-3\text{ dB}\right|<20\text{ dB}. \quad (7)$$

When pseudolite signal pulse duration equal to 2 ms and interval accumulation of the correlator equal to 10 ms then allowable power range boundaries pseudolite radiation signal depicted in Fig. 2.

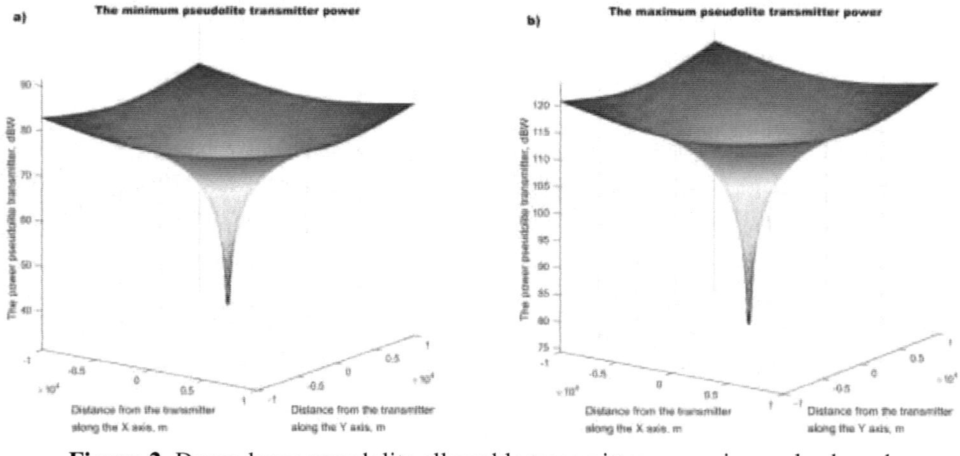

Figure 2. Dependence pseudolite allowable transmitter power in a pulsed mode of operation of the distance in the plane of the Earth's surface. Used Hanning window, parameter a = 1: *a* – minimum power value; *b* – maximum power value.

IOP Conf. Series: Materials Science and Engineering **255** (2017) 012010 doi:10.1088/1757-899X/255/1/012010

Despite the fact that the calculated output power of the pulse signal can have values pseudolite 40 dBW above, the actual radiation power permissible dynamic range is limited to the automatic gain control and analog-to-digital converter of the receiver. To ensure isolation of pseudolites and GLONASS signals taken receiver on the same frequency using only code division will not be enough, as the GLONASS signal will be simply muted by pseudolite signal. At the same time ensuring reliable receive signal on the same channel and knowing pseudolite pulse signal width we can compensate pseudolite signal in the second channel on the same frequency.

Thus, the first correlator will collect samples over the entire range of the pseudolite signal accumulation interval T, the second correlator in the interval $T - \tau_p$ accumulation will collect a signal of the GLONASS navigation system.

3. Using the weighting windows to improve the coherence of signals pseudolites and navigation spacecraft

The disadvantage of the above approach is the influence of the side lobes of the spectral characteristics of a pulsed signal pseudolite to adjacent frequency channels of GLONASS. To improve the frequency response of the working channel in consumer navigation equipment can use weighting window.

So, for the Hanning window with parameter value $\alpha = 1$, the level of the first side lobe is –23 dB, which corresponds to the level of interference from adjacent frequency channels of the receiver. Using Hanning window function with parameter a = 4 to forming a pulse pseudolite provides a side-lobe level –46.8 dB.

To estimate the allowable power of the radiation pulse pseudolite range using this Hanning window function can estimate the signal at the output of the consumer navigation equipment correlator. Let us assume that the receiver does not have Doppler frequency shift and the received signal is completely correlated with the code sequence

$$I_{i,PL}(t) = \int_0^{\tau_p} h_{i,DC}(t) h_{i,PL}(t-\tau) \frac{\left(3 + 4\cos(2\pi t) + \cos(4\pi t)\right)}{8} \cos\left(\Delta\omega_{i,DC} t + \Delta\varphi_{i,DC}\right) dt \cdot \quad (8)$$

4. Conclusion

An analysis of pseudolites permissible power range modeling results allows us to conclude that when pseudolites are used together with the GLONASS system, the impulse mode of operation of pseudolites is most appropriate.

When a pseudolite is pulsed possible to implement time division of the GLONASS signals and pseudolites signals received by one frequency channel on the basis of data on the pulse duration. Since the "near-far" effect is arises due to difference between the power of the received GLONASS signals and pseudolites signals, to reduce the interference pulses adjacent channels on each other, the pulses should be formed through the weighting windows, side lobes which have a level below the minimum level of signal reception pseudolite.

Fig. 3 shows that the dynamic range of operation of the receiving channel without the appearance of the effect of "near-far" should be about 40 dB. Using a Hanning window with parameter a = 4 pulse weighing provides a side-lobe level of 6 dB below the minimum power level pseudolite pulse signal, thus providing isolation adjacent frequency channels.

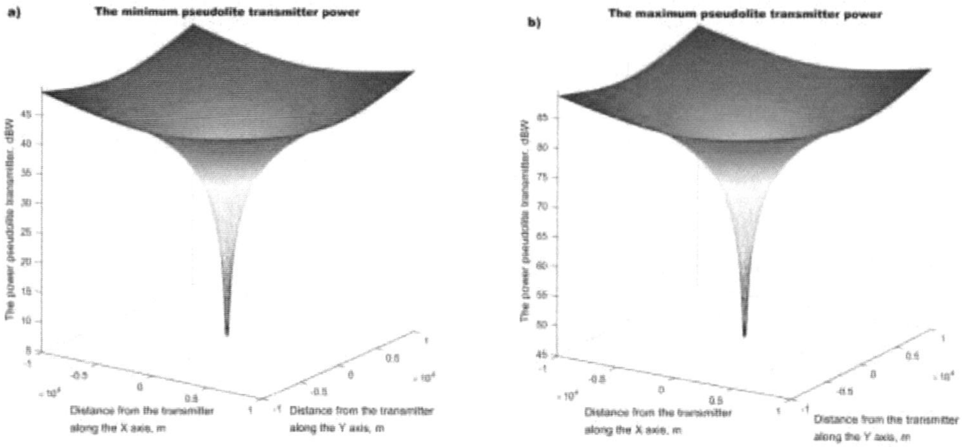

Figure 3. Dependence pseudolite allowable transmitter power in a pulsed mode of operation of the distance in the plane of the Earth's surface. Used Hanning window, parameter a = 4: *a* – minimum power value; *b* – maximum power value

The research had been conducted with financial support from the Russian Science Foundation. Grant No 16-19-10089.

References

[1] Kim C, So H, Lee T, Kee C. 2014 A pseudolite-based positioning system for legacy GNSS receivers *Sensors (Switzerland)* vol **14 (4)** pp. 6104-6123

[2] Xu R, Chen W, Xu Y, Ji S 2015 A new indoor positioning system architecture using gps signals *Sensors (Switzerland)* vol **15 (5)** pp 10074-10087.

[3] Fateev Yu L, Ratuschnyak V N, Kartsan I N, Tyapkin V N, Dmitriev D D, Goncharov A E 2016 Analyzing measurement errors for navigation parameters in onground short-range navigation systems based on pseudolites *IOP Conf. Series: Materials Science and Engineering* vol **155**

[4] Liu C, Lin P, Zhang J, Zhao X 2015 An improved pseudolite location model of GPS/pseudolite integrated positioning *Zhongguo Kongjian Kexue Jishu/Chinese Space Science and Technology* vol **35 (1)** pp 75-83

[5] Van Dierendonck A J, Fenton Pat, Hegarty Chris 1997 Proposed airport pseudolite signal specification for GPS precision approach local area augmentation systems *Proceedings of ION GPS* vol **2** pp 1603-1612

[6] Li C, Chen J, Chu H 2011 Design of the pulse duty cycle of the pseudolite signal in local area augmentation systems *Science China: Physics, Mechanics and Astronomy* vol **54** pp 1004-1008

[7] Dvorkovich V P, Dvorkovich A V 2016 Window functions for a harmonic signal analysis. 2rd ed. (Moscow: Tekhnosfera)

Pseudolite systems for close-range navigation: the problem of synchronization*

I N Kartsan[1], A E Goncharov[2], D D Dmitriev[3], V N Ratuschnyak[4], I V Kovalev[5]

[1]Associate professor, Reshetnev Siberian State University of Science and Technology, Krasnoyarsk, Russia
[2]Associate professor, Reshetnev Siberian State University of Science and Technology, Krasnoyarsk, Russia
[3]Associate professor, Siberian Federal University, Krasnoyarsk, Russia
[4]Associate professor, Siberian Federal University, Krasnoyarsk, Russia
[5]Professor, Reshetnev Siberian State University of Science and Technology, Krasnoyarsk, Russia

E-mail: kartsan2003@mail.ru

Abstract. This paper discusses the implementation of an electronic pseudolite close-range navigation system provided that one of the pseudolites is located at a certain altitude. The authors have developed principles for building a timescale for this system to solve the problem of mutual synchronization of the pseudolites. The principle of retransmitting the signals from all the pseudolites to one reference station is considered as the basis for this system.

1. Introduction

Global navigation satellite systems (GNSS) have become the most widespread navigation systems enabling to accurately determine an object's position, its speed, and local time. On the basis of the main functions and concepts of GNSS, a multitude of secondary services and systems has been designed, including local and regional differential subsystems, systems for monitoring and controlling various types of vehicles, systems for monitoring and tracking cargos and objects, universal time-management subsystems, etc. For some applications, GNSS signals may be used as reference signals.

Despite their enormous popularity, GNSS are seldom applied for autonomous navigation of aircraft in the airspace (traffic zones) of aerodromes; unmanned flight control; navigation of vessels in harbors, narrow water strips, and rivers; providing information support for geodic and cartographic works. One reason for the limited application of GNSS in the mentioned fields is their low immunity to interference. This factor makes it difficult to cope with the requirements of continuity and accuracy for navigation systems.

A possible solution to this problem is the addition of a close-range ground radio-navigation system to the GNSS. Using ground-based pseudolites, the ground system will form a navigation field in a given area [1]. A high level of accuracy in such navigation systems is provided by accurate geodic ties and a mutual synchronization of the pseudolites, which are

* The work was supported by the Russian Science Foundation (Grant No. 16-19-10089).

stationary navigation signal transceivers. These signals are identical to the ones emitted by GLONASS satellites. The aspects of using pseudolites for GPS had been previously intensely discussed in [2].

The main problem in building a navigation system is synchronizing the pseudolites and the navigation devices of the user. As discussed in [3; 4], it is necessary to position the pseudolite at a specific altitude (elevation) to reduce the value of the geometric factor in terms of height. With this done, the measurement accuracy for the altitude of a moving object is improved. However, this results in problems with synchronizing and affects the measurement accuracy for the coordinates of the pseudolite, which is situated on a balloon-based system. A possible solution lies in the retransmission (re-radiating) of navigation signals from all the pseudolites to one reference station (Fig. 1).

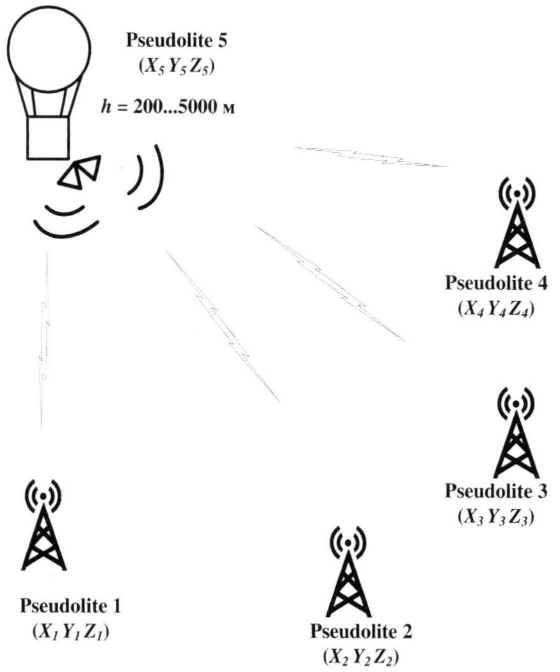

Figure 1. demonstrates the remote determination of the coordinates and speed of a mobile balloon-based pseudolite

2. A discussion on the approaches to solving the problem of mutual synchronization

Retransmitting signals via the suspended pseudolite may be used for the following:

1. Remote determination of the coordinates and speed of a mobile balloon-based pseudolite.

2. The determination of the coordinates of a pseudolite, which fails to receive signals from satellites (or is receiving them on a small scale, failing to fulfill its navigation and time tasks) by means of retransmitted pseudolite signals, emitted by a ground-based pseudolite.

Determining an object's navigational parameters (coordinates, speed and direction of movement, time and frequency scale, spatial orientation) remotely is possible after solving the navigation and timing problem, utilizing the measurement results for the radio-navigation parameters (delays, Doppler shift, etc.) of the pseudolite retransmission systems in receivers and processing units of retransmitted pseudolite signals as follows:

$$\begin{cases} R_1^2 = (x_5^{'} - x_1)^2 + (y_5^{'} - y_1)^2 + (z_5^{'} - z_1)^2 = C^2(\tau_5^{'} + t_1)^2 \\ R_2^2 = (x_5^{'} - x_2)^2 + (y_5^{'} - y_2)^2 + (z_5^{'} - z_2)^2 = C^2(\tau_5^{'} + t_2)^2 \\ R_3^2 = (x_5^{'} - x_3)^2 + (y_5^{'} - y_3)^2 + (z_5^{'} - z_3)^2 = C^2(\tau_5^{'} + t_3)^2 \\ R_4^2 = (x_5^{'} - x_4)^2 + (y_5^{'} - y_4)^2 + (z_5^{'} - z_4)^2 = C^2(\tau_5^{'} + t_4)^2 \end{cases}$$

This approach may be used as an alternative to the traditional system of determining an object's navigational parameters remotely, built on the basis of a set of standard radio-navigation hardware, which provides navigation and timing services, and signal transmitting hardware. In a situation when retransmission is utilized, the information transmitting devices are replaced by retransmitting devices; in this case, the radio-navigation equipment becomes lighter, less bulky, more energy-efficient, and, as a result, less expensive. This is achieved by eliminating such nodes as the digital signal processing unit and the computing unit.

An additional advantage to executing navigation and timing services on retransmitted signals is the relative mode of determining the real-time navigational parameters of an object using receiving and processing devices for retransmitted signals. For this mode to function, it is not necessary to transmit any supplementary data from the object. Thus, since the solution of the navigation and timing problem is found in the receiving and processing devices for retransmitted signals, the object itself does not contain data on its navigational parameters.

During the distribution of the pseudolite signal, an additional delay occurs along its retransmission path, which is determined by the distance between the GNSS repeater and the retransmitted signal receiver and processing unit. Another Doppler shift also occurs, as there is a mutual motion of the repeater and the retransmitted signal receiver and processing unit. These factors alter the results of measuring the radio-navigational parameters of the pseudolite, which are executed in the retransmitted signal receiver and processing unit, with the objective of providing navigation and timing services, as well as determining the navigational parameters of the balloon-based pseudolite. An additional factor that alters the results of the decisions made by the receiver and processing unit of the retransmitted signals for solving the navigation and timing problem for balloon-based pseudolites is the deviation of the frequency and time scale of the receiver and processing unit from the frequency and time scale of the balloon-based pseudolite.

In order to eliminate the influence of the distribution path on the solution of the navigation and timing problem for the receiver and processing unit of the retransmitted signal, as well as the parameters of the frequency and time scale of the receiver and processing unit of the retransmitted signal, a special pilot signal, generated in the repeater and transmitted from the object with the pseudolite retransmitted signal, can be implemented. Such parameters as delays and the Doppler shift of the pilot signal in reference to the frequency and time scale of the retransmitted signal receiver and processing unit are measured by the reference station. These results determine the additional pseudolite signal delay, which is dictated by the retransmitted signal path and deviations from the timescale of the receiver and processing unit, together with an additional Doppler shift of the pseudolite signals. The Doppler shift is caused by the mutual motion of the repeater and the receiver and processing unit of the retransmitted signal, along with the frequency deviation of the reference generator of the receiver and processing unit of the retransmitted signal from the reference signal generator of the repeater.

There are a number of options on utilizing the pilot signal. The first option supposes that the reference signal generator of the receiver and the processing unit of the retransmitted signal can be synchronized by the pilot signal. In this variant, an automatic frequency tuning counter (analog or digital) of the reference signal generator of the receiver and processing unit of the retransmitted signal, adjusting to the frequency of the received pilot signal, is used. One advantage of this method is that there is no need for any additional complicated software for the receiver and processing unit of the retransmitted signal during the initial stage of processing and solving navigation and timing problems. A disadvantage to this method, which utilized a direct automatic frequency tuning of the reference signal generator of the receiver and processing unit of the retransmitted signal, adjusted to the frequency of the pilot signal, is a limit to the system's bandwidth capabilities. In this case, the receiver and processing unit can processes only one signal from a repeater at a time.

Another option for utilizing the pilot signal is the measurement of its frequency regarding the reference signal generator of the receiver and the processing unit of the retransmitted signal. In this case, the synchronization of the processes in the receiver and the processing unit of the retransmitted signal is executed from a specific reference signal generator, the frequency of which is not tuned to the frequency of the received pilot signal. The receiver and processing unit of the retransmitted signal assesses the frequency deviation of the pilot signal from a nominal value. Using this data, an additional component, determined by the distribution path and the parameters of the frequency and time scale of the receiver and processing unit of the retransmitted signal, is excluded from the frequency assessment of the pseudolite retransmitted signal.

A third option is more complicated in terms of its implementation as it requires adjusting the initial data processing software for the receiver and processing unit of the retransmitted signal. A positive aspect is the absence of any principal limitations for the bandwidth capabilities of the receiver and processing unit of the retransmitted signal. One device can service a few repeaters.

The receiver and processing unit of the retransmitted signal can operate in such a mode in which the navigation and timing problem is solved as if it were taking place on a balloon-based pseudolite. The receiver and processing unit of the retransmitted signal is, in its way, a structural element of the navigation receiver of an object, e.g. the digital signal processing unit and the computing unit. The repeater, the retransmission path, the receivers and processing units of the retransmitted signal are included in the discontinuity between the reception antenna, the signal path (a section), and the remaining units of the traditional radio-navigation system. In this mode, the receiver and processing unit of the retransmitted signal, along with the reception of retransmitted signals from the repeater in a retransmission frequency range, executes the reception of pseudolite signals in a standard frequency range.

3. Conclusion

The mutual processing of the measurement results of the pseudolite signal, received by the navigation antenna receiver and processing unit of the retransmitted signal, and the pseudolite signal, received by the object and retransmitted by a repeater, allows one to implement a highly-accurate determination of the object's coordinates in respect of the receiver and processing unit of the retransmitted signal. In this mode it is possible to implement a code and phase mode for determining the relative coordinates [5]. The phase mode can be implemented due to the fact that the receiver and processing unit of the retransmitted signal, on account of the pilot signal, contains all the necessary data on the frequency and time scale of the object.

Thus, we have discussed on a number of approaches to solving the problem of synchronization in pseudolite navigation systems. Further inquiry into this problem will provide an answer to the topical GNSS issues, particularly in relation to GLONASS.

References

[1] Perov A I, Kharisov V N (eds.) 2010 GLONASS. Printsipy postroeniia i funktsionirovaiia [GLONASS. Basic concepts and operation (in Russ.)] 4th ed. (Moscow: Radiotekhnika)
[2] Steward Coob H 1997 GPS pseudolites; theory, design, and applications (Ph.D. dissertation). Department of Aeronautics and Astronautics and the Committee of Graduate Students. Stanford University. H. Steward Cobb.
[3] Fateev Yu L, Ratuschnyak V N, Kartsan I N, Tyapkin V N, Dmitriev D D, Goncharov A E 2016 Analyzing measurement errors for navigation parameters in onground short-range navigation systems based on pseudolites *IOP Conf. Series: Materials Science and Engineering* **155**
[4] Tyapkin V N, Garin E N, Dmitriev D D, Ratushniak, V N (2016) Study characteristics of coordinate measurement in the ground system of navigation based pseudosatellites, *Uspekhi sovremennoi radioelektroniki*, no. 11 P. 132–136.
[5] Garin E N, Dmitriev D D, Kokorin V I, Kremez N S 2006 Opredelenie otnopsitel'nykh koordinat ob"ekta s pomoshch'iu sputnikovykh sistem radionavigatsii [Determining the relative coordinates of objects using satellite radio-navigation systems (in Russ.)] *Radiolokatsiia, navigatsiia i sviaz': conference proceedings "RLNC-2006"* Voronezh vol **3** pp 1776–1786

IOP Conf. Series: Materials Science and Engineering **255** (2017) 012012 doi:10.1088/1757-899X/255/1/012012

Theoretical and experimental research of organic Rankine cycle steam turbine plants

A A Kishkin, A V Delkov and M G Melkozerov
Reshetnev Siberian State University of Science and Technology,
31 Krasnoyarskiy Rabochiy av., Krasnoyarsk, 660037, Russia.

E-mail: delkov-mx01@mail.ru

Abstract. Currently steam power cycles using organic actuation fluid - Freon, ammonia, ethanol, isobutene, etc are becoming increasingly important. Such cycles are called Organic Rankine Cycle (ORC). With the help of such cycles it is possible to use low-grade heat sources in the production of mechanical and electrical energy.

The modern energy crisis is problem area in the supply of energy resources to an economy. This calls for giving more emphasis on the development of alternative energy sources. One of the common approaches to obtain the mechanical and electrical energy is the application of the heat engines.

According to the analysis of domestic and foreign publications it is possible to indicate four types of heat sources that are promising for further development and they require to design a specific approach to their use as energy resource [1-4]:

- geothermal heat;
- thermal output of solar radiation;
- thermal industrial emissions;
- heat flow from the engines and airborne equipment transport systems, including vehicles, marine transportation, spacecraft.

The particular feature of the above sources means the presence of non-ambient temperature that is not sufficient to organize the traditional steam cycles in water vapor. Temperature difference for these sources, the difference in temperature between the heat source and the environment, is in the range 60-200 C. In the papers, such sources are called low-temperature [5-6].

The most promising and the most common methods to transform thermal energy into mechanical and electrical ones are the steam power cycle in turbo machines. Such cycles function on the principle of heat transfer from the source to the field of warm reset producing the energy. At these temperature differences steam power cycle can be realized only through specific actuation fluids - organic (ethanol, Freon, isobutene, etc.) [7].

The research and development of energy resources are important because there are still no adequate and optimal methods for their use.

Engineering development of these steam turbines plants (STP) on organic acuation fluids (OAF) presented on the world market (Turboden (Italy), Infinity Turbine LLC (USA)), have effective coefficiency in the range of 10-12% and limited application spheres. Powerful methods for design and optimization of such systems have not been developed yet.

Connecting with the above material the need for analysis, calculation and design of such plants appear. The urgency of the work to create steam turbines plants on organic actuation fluids causes the need to simulate cycles to describe the plant and optimize its processes [8].

Creation of mathematical models is a challenging direction in modern research [9-11]. The model allows calculating the main parameters of the process under certain initial conditions, to get them modified by varying the input data, to assess the impact of various factors on the plant functioning.

The mathematical model defines the joint work of constituent elements of stream turbine plants on organic actuation fluids. Block diagram of the plant (Fig. 1) includes the following elements:

1. Turbine: an active axial turbine is used to convert the energy of the actuation fluid into the work.
2. Circulation pump: it is designed to increase the pressure of the actuation fluid and feed it into the evaporator.
3. Evaporator: it is assigned to transfer heat from the source to the actuation fluid.
4. Capacitor: it is used to transfer the heat of the actuation fluid source to cold working fluid and to change the actuation fluid to the liquid phase.
5. Electricity generator.

The principle of operation of the plant is the following. A working body receives heat from a source in the evaporator, where it is vaporized and heated. After that, the actuation fluid enters the turbine where it emits energy. Pressure and temperature of the actuation fluid is reduced at the same time. Further, the working fluid enters the condenser, where it is condensed due to the interaction with the source of the cold. The liquid actuation fluid goes to the receiver, after it is evacuated by the circulating pump and is fed to the condenser.

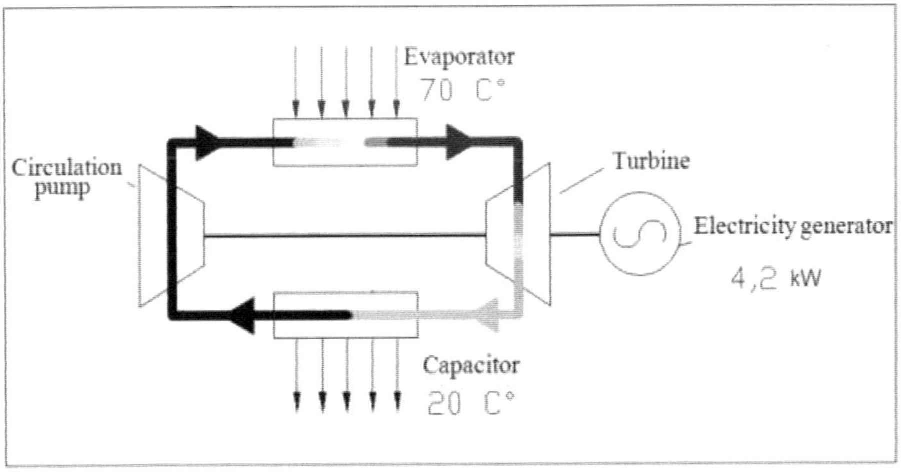

Figure 1. Block diagram of the plant

The model plant incorporates mathematical models of its constituent elements to define the basic parameters of the actuation fluid that allow estimating the reaction of equipment to both internal and external influencing factors. In the model plant there are some main parameters to provide the required mode of operation such as: pressure p, consumption V, temperature T and angular speed of turbine rotor motion w.

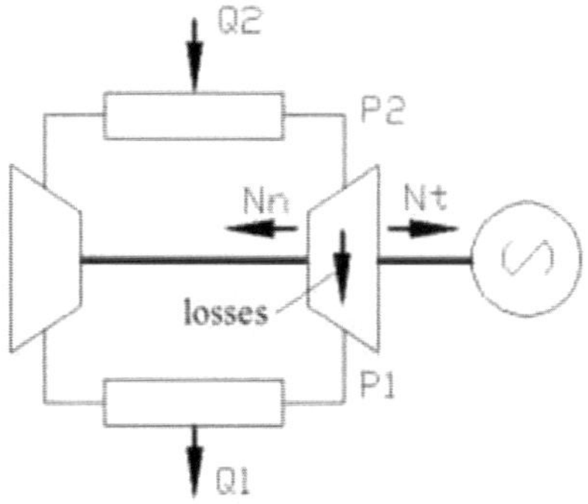

Figure 2. The energy balance of the system.

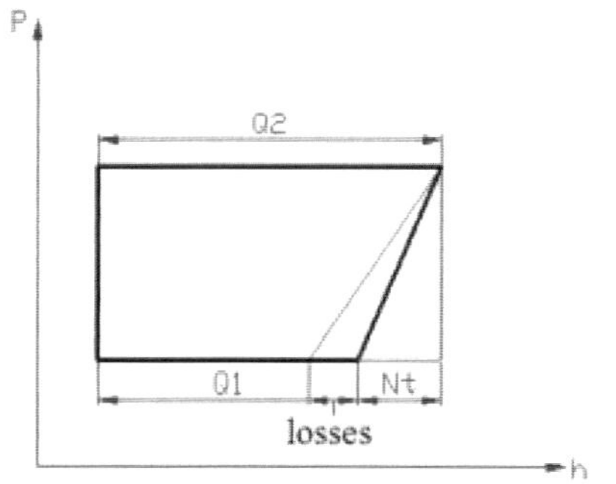

Figure 3. Diagram of the working cycle of the plant.

The energy balance of the model is as follows (Fig. 2). In the adiabatic turbine power N_{ad} is produced, that depends on the parameters of the actuation fluid at the inlet and outlet position and the mass flow rate.

$$N_{ad} = L_{ad} \cdot \dot{m} = \frac{k}{k-1} RT_0 \left[1 - \left(\frac{P_2}{P_1} \right)^{(k-1)/k} \right] \cdot \dot{m} \tag{1}$$

where L_{ad} - adiabatic work of gas; $\dot{m}$ - gas flow in the turbine; k - rate of adiabatic line; P_2 - gas pressure at the turbine inlet; P_1 - gas pressure at the turbine outlet;

Adiabatic power is used for the technical capacity of the generator *Nt*, pump power *Nn* and losses in the turbine itself N_{loss}.

$$N_{ad} = Nn + Nt + N_{loss} \tag{2}$$

The balance of power can be seen in the diagram of the working cycle (Fig. 3).

$$Q_2 - Q_1 - (Nn + Nt + N_{loss}) = 0 \tag{3}$$

where Q_1 - Heat flow in the condenser; Q_2 - Heat flow in the evaporator.

As the all parameters are determined based on the actuation fluid flow in the system, respectively it sets the condition of continuity in the system.

In addition to establishing energy balance and expenditure, the system determines the angular velocities of the components – rotors of turbines, pumps, generators. Angular velocity determines the disk friction losses, mechanical losses, etc. In general, the power generator and pump losses in the turbine can be introduced with a function of angular velocity, the following equation is possible to obtain

$$N_{ad}(P_1, P_2, T_0) = Nn(\omega) + Nt(\omega) + N_{loss}(\omega) \tag{4}$$

This condition furnishes the angular velocity of the system.

The model allows calculations in two directions, solving forward and inverse design tasks.

The forward task is to design a steam turbine, if you know the input parameters of evaporator and condenser.

Initial data for the forward problem are the characteristics of the source; characteristics of the refrigerator; useful power of turbine; rotor angular velocity; organic actuation fluid (Freon) and its chart.

The solution of the forward task is the following algorithm:

1. The working cycle of plant is created based on the known parameters of the source and the refrigerator; its specific parameters are defined.
2. The specific parameters of the turbine are calculated: the adiabatic work L_{ad}, useful work Ln, losses and efficiency.
3. The specific parameters of the pump are determined: expended power N_z, useful power Nn, losses and efficiency.
4. The specific parameters of heat exchangers and their geometry are defined.
5. The mass flow output at the evaporator based on the known thermal power source Qist, mass and energy parameters of the cycle are calculated.
6. The type and geometry of the turbine, pumps, heat exchangers are selected.

The result of solving the forward task is defined by the geometry of the device, its energy balance, the power output and efficiency.

The inverse task is to obtain parameters of different modes, features of the stream turbine plant if its geometry is known, then it is possible to optimize cycle of this plant based on the parameters obtained.

Initial data for the inverse task are the geometry of the turbine, pump, heat exchanger, the actuation fluid and its properties; angular velocity of motion of the rotor, the characteristics of the source and refrigerator, range and a step change in controlling parameters.

The solution of the inverse problem requires the following algorithm: the adiabatic gas work and loss in the nozzle unit is determined according to the given pressure and mass flow at the inlet of the nozzle unit. Next for the Freon flow in the impeller channels of the rotor wheel the velocity triangles are constructed, the velocity at the outlet of the working grid and losses are defined. Taking into account the pressure and temperature at the outlet of the turbine together with the characteristics of the refrigerator the parameters in the condenser are evaluated. The pumping capacity and its efficiency is measured based on the flow rate and pressure drop.

The result of solving the inverse problem is sets of design parameters for different combinations of controlling factors which are the basis to determine optimal modes of the plant operation and critical situations.

The inverse problem can be solved only by taking into consideration the actual losses in the plant, which are in most cases found experimentally. Loss imbalance is necessary to determine the components' effectiveness.

Determination of losses due to the complexity of their analysis is possible only by experiment. The algorithm if there is a shortage of experimental data is designed to meet the recommended values of loss coefficients for typical steam turbines plants: in the turbine nozzle apparatus φ; in the rotor ψ. Losses with output speeds are determined by the triangle of velocities. Ventilation losses and losses from the disk friction are found out with the help of friction in the spatial boundary layer. Mechanical losses are determined with the speed and pressure in the system.

IOP Conf. Series: Materials Science and Engineering **255** (2017) 012012 doi:10.1088/1757-899X/255/1/012012

Based on the above equations of the mathematical model of the STP OAF the calculation algorithm was set up and the numerical study of the operating modes of the plant and energy balance of the turbine was carried out. The variable parameter is pump thrust that gives the pressure difference between the heat exchangers. In this system two parameters are changed, they are specific work and the mass flow rate.

Calculation algorithm is characterized by the following parameters: the original data are the heater temperature, the refrigerator temperature, the working elements geometry (profile of nozzle block, the area of heat exchange evaporator and condenser, the diameter of the turbine wheel, the typical angles of the nozzles, inlet and outlet vanes). The calculated parameters are mass flow rate of the actuation fluid, adiabatic turbine capacity, the losses in the turbine, the thermodynamic parameters of the actuation fluid at the inlet and outlet of the turbine, the speed of the actuation fluid in the turbine wheel, the required pump power, the efficiency of the turbine.

The solution is as follows: the basic thermodynamic parameters of the cycle - pressures and flows, temperatures of boiling and condensation are determined the original data. Further, the parameters of the turbine are calculated. The adiabatic operation of gas and loss in the nozzle are determined with the calculated pressure and specific volume at the inlet of the nozzle unit. The velocity triangles are designed the velocity at the outlet of the working grid, the loss of a wheel and the output speed are determined for steam in the impeller channels of the rotor wheels. The angular velocity and mechanical ventilation losses are evaluated next. To calculate them the parameters of the working body are taken from a database of properties that represent a mathematical surface in a formal way.

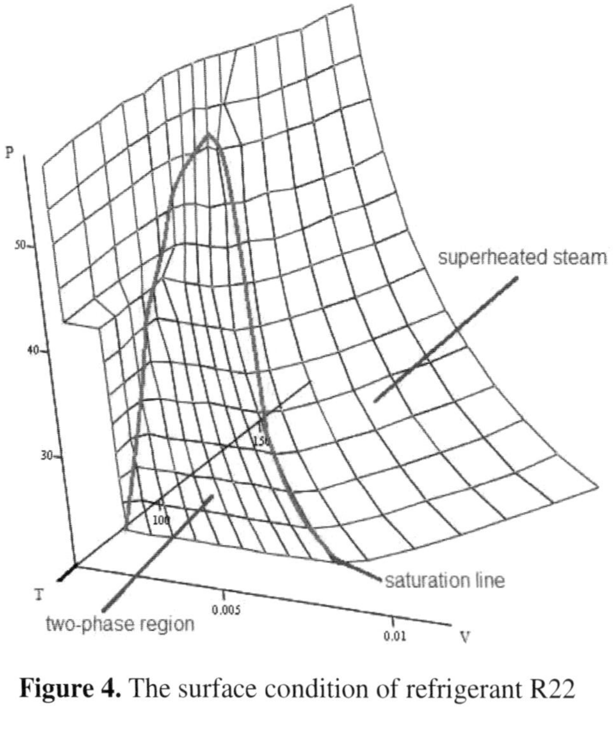

Figure 4. The surface condition of refrigerant R22

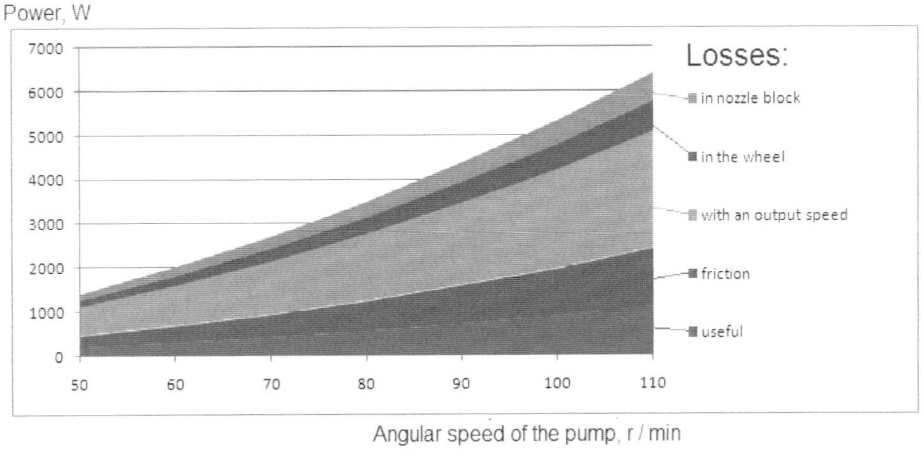

Figure 5. The balance of power turbine

The mathematical surface of the state (Fig. 4) of low-temperature boiling actuation fluid within the pressure P, specific volume V and temperature T allows calculating cycle power plants by numerical methods taking into account the continuous changes in the properties. In addition, using the surface it is possible to obtain the basic parameters of the body (enthalpy, entropy, heat capacity, sound velocity, etc.) applying the differential equations of thermodynamics. Representation of the surface is constructed on the state of refrigerant tables.

The results of calculation of the turbine balance are based on the model shown in Fig. 5. There is a tendency to increase the efficiency with the increased power adiabatic turbine. This is due to rearrangement of the velocity triangles and the decrease of losses with the output speed. Optimal solutions for the efficiency are obtained with greater pump thrust.

To verify the model test setup was designed. Measurement of the effective turbine capacity is done with pneumothermometric method on the consumer that is hydraulic brake. A centrifugal pump pumping the water is used as an alternative to the hydraulic brake. To assess the turbine capacity the parameters of energy imparted to the water are required, which is estimated by the data connected with pressure and water temperature at the inlet and outlet of the pump.

The main element of the laboratory installation (Fig. 6) is the steam turbine. Such autonomous turbines are always performed active, with a supersonic gas flow. Supply of the actuation fluid to the turbine blades is through a single profiled nozzle.

Experimental studies in the pilot plant, processing and analysis of experimental results and model verification during the experiment have not been done by the authors yet.

In addition, the revision of the mathematical model of steam turbine is being planned. The coefficients of estimated losses are going to be corrected; they are determined by the analysis of experimental data. As a result, on the basis of the model it is intended to optimize steam turbine.

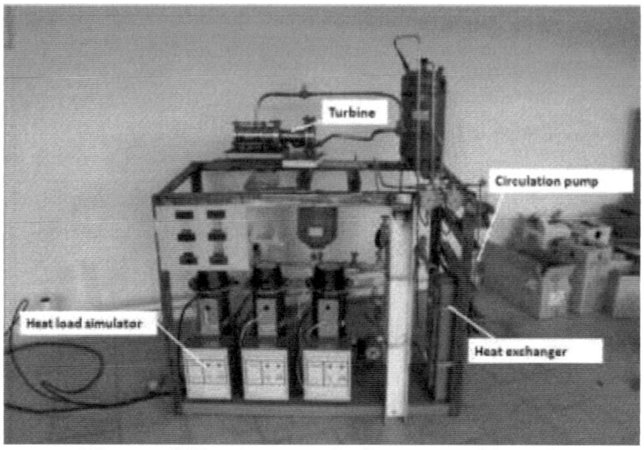

Figure 6. Testing stand of steam turbine plant.

Optimization of the organic steam turbine cycle can be based on a mathematical model using the energy equation and it describes the system components parameters. Depending on the type of problem being solved the model allows calculations in two directions, they are the definition of the geometry of the plant elements according to the given parameters of operating temperatures and the calculation of the energy parameters of the plant based on the given geometry. A mathematical model is an effective tool for the design of such plants.

References

[1] Brasz Joost J., Biederman Bruce P., Holdmann Gwen 2005 Power Production from a Moderate Temperature Geothermal Resource *Geothermal Resources Council Annual Meeting* (Reno, NV, USA)
[2] Karellas S, Schuster A 2008 Supercritical Fluid Parameters in Organic Rankine Cycle Applications *Int. J. of Thermodynamics* vol **11** (No 3) pp 101-108
[3] Glavatskaya Y, Olivier G, Shonda O F, Podevin P 2011 Heat recovery systems for passengers vehicles, *"Buletinul Ştiinţific al Universităţii din Piteşti, seria Autovehicule Rutiere"* No **21** **(2)** pp 93-104
[4] Hatchman J C 1991 Steam cycles for waste heat recovery: a case study N&O Joernaal September pp 32-38
[5] Quoilin S, Lemort V, Lebrun J 2010 Experimental study and modeling of organic Rankine cycle using scroll expander, *Applied energy* **87** pp 1260-1268
[6] Canada S, Cohen G, Cable R, Brosseau D and Price H 2004 Parabolic Trough Organic Rankine Cycle Solar Power Plant, *DOE Solar Energy Technologies Program Review Meeting* (Denver)
[7] Hung T C, Shai T Y, Wang S K 1996 A review of organic Rankine cycles for the recovery of lowgrade waste heat (Kaohsiung Polytechnic Institute, Taiwan)
[8] Madhawa H D Hettiarachchia, Mihajlo Golubovica, William M. Woreka, Yasuyuki Ikegamib 2007 Optimum design criteria for an Organic Rankine cycle using lowtemperature geothermal heat sources (Chicago)
[9] Samarskii A A, Mikhailov A P 2001 Mathematical modeling: Ideas. Methods. Examples. (Moscow: Fizmatlit)
[10] Popyrin L S 1978 Mathematical modeling and heat power plant optimization. (Moscow: Energiya) p 416
[11] Semenov A G 2003 Mathematical models in engineering. (Kemerovo) p 96

IOP Conf. Series: Materials Science and Engineering **255** (2017) 012013 doi:10.1088/1757-899X/255/1/012013

Investigation of swirling diphasic flow in centrifugal phase separator

A A Kishkin[1], M G Melkozerov[1], A V Delkov[1]
Reshetnev Siberian State University of Science and Technology,
31 Krasnoyarskiy Rabochiy av., Krasnoyarsk, 660037, Russia

E-mail: delkov-mx01@mail.ru

Abstract. Separation of disphasic mediums form the basis of many technological processes. However calculation methods for diphasic mediums used in industry found on empirical studies and cannot lay a claim to universality. In the offered model of swirling diphasic flow the flow is divided into the viscous interface and the diphasic core. To research the flow of liquid in the area of the viscous interface with big gradients of velocity Navier-Stokes equations are used.

The analysis of the existing methods to separate gas-liquid mixtures shows that in the basis of phase separation the density heterogeneousness of the diphasic flow is used efficiently.

In the constant gravity environment the density heterogeneousness of the resting liquid is the source of Archimedean force influencing particularly gas bubbles and making them float to the free surface.

If body forces spring up as the result of the liquid movement, then their values and directions depend upon distribution of the velocities in the flow and eventually are defined by pressure forces and viscosity forces. In such an environment with the liquid movement in curvilinear channels and with flow swirling caused by the swirler in the tubes internal inertial forces spring up.

Otherwise inertial forces in moving systems (channels) spring up. In this case inertial forces are external to the moving flow and will be defined by the environment of the system movement not by the flow. Inertial forces springing up with the system spinning in its accelerated or slowed-down movement would belong to this category.

The usage of centrifugal inertial forces phase separating devices implies creation of the area with the swirled sheeted diphasic flow. With all that the centrifugal force field can be both external and internal to the flow. In the first case the body forces will be defined by the parameters of the spinning system concerned with the flow and entirely depend upon it. In the second case the level of inertial forces will definitely concern with the forces causing forced movement.

The first variant is realized in the devices of rotary type that have mechanical drive with the available power of the shaft. In the second case the spin is carried out by special devices that can be subdivided into three types: screw inserts, tangential channels and spade swirlers.

In different fields of industry centrifugal separators (phase separators, cyclones, hydrocyclones) are applied to separate gas-liquid systems. They are remarkable for simplicity of construction, great carrying capacity, low materials consumption, being easy in use and high performance.

IOP Conf. Series: Materials Science and Engineering **255** (2017) 012013 doi:10.1088/1757-899X/255/1/012013

The mechanical method of phase separation is most economizing particularly in degasation in the field of centrifugal forces. In this connection the devices of hydrocyclone type have recently got their new application. They have come to be used to refine liquids from dissolved and dispersed gases. With the rise of technological demands of gas content of liquids new hydrocyclone constructions have been designed. They make it possible to provide effective phase separation in each definite case.

Centrifugal separation of gas-liquid mediums is remarkable for high performance and is widely used in heat exchange and mass transfer equipment. Centrifugal phase separator of various types have become widely spread.

Schematically the construction of centrifugal phase separator (fig. 1) can be presented in the following details:

- diphasic mixture supply (1);
- device to transform entrance axial flow into the swirled one (the swirler) (2);
- swirling chamber (3);
- circumferential pipe-bend of the gas phase (4);
- central (axial) pipe-bend of the gas phase (5).

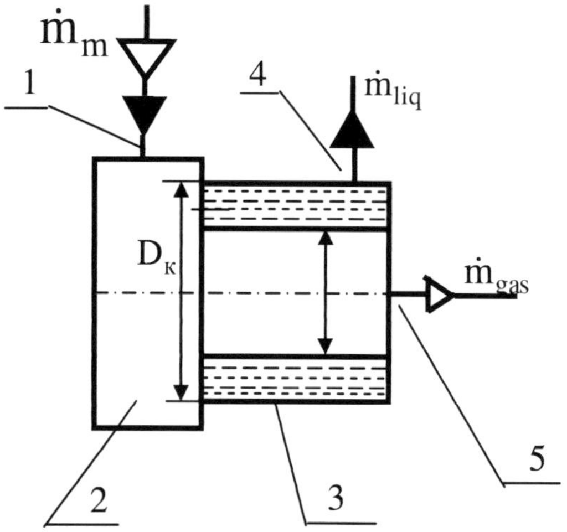

Figure 1. Centrifugal phase separator

Separation of diphasic flow is based on the fact that in the centrifugal field there is separation force that influences gas insertions in the liquid and shifts them to the axis of the chamber. In the centre of the chamber there springs up a gas vortex squeezed by the liquid ring. The separated swirled flow is moving towards the axis, at the outlet of the chamber diverting of separated phases is carried out.

Pilot researches of degasation of gassy liquid in the phase separator showed that velocity enhancing of the liquid flow in the inlet of the device ambiguously influences efficiency of the device. For gassy liquids velocity enhancing raises efficiency but within the definite scope. Getting some certain value of the liquid velocity in the inlet of the hydrocyclone increases drag-out of gas bubbles in outgoing flow of the liquid. In this case gas bubbles do not manage to enter the central zone of the phase separator.

Stability of gas-liquid mixture is very largely defined by dispersity of the gas in the liquid. However even with high gas dispersity noticeable phase separation happens in a fraction of a minute.

The main factors to define characteristics for the course of the swirled flow are:

- geometry of continuous-flow part of the chamber in the phase separator;
- regime parameters (consumption of operational liquid and gas);
- characteristics of operational liquid (density, viscosity, temperature);
- characteristics of gas (density, temperature);

The valid obstacle in projecting is the fact that centrifugal phase separators in spite of being widely spread are still very little studied both experimentally and theoretically. It is explained by complexity of running fine model researches at pilot plants.

Lack of information about the processes that really take place during separation of diphasic mixtures blocks development of theory and engineering designs and can be compensated by proper analysis of hydrodynamics equations of swirled flows and designing on its basis a mathematical model for the process of mixtures phase separation.

The basis of diphasic flow theory exists in classic mechanics of liquid and gas and explains in details movement of each phase. However its up-to-date condition is so that applying strict conclusions from fundamental correlations such as Navier-Stokes equations is not possible because calculations of diphasic flows are not reliable. The combination of the factors that cause all this is the following: endless variety of geometric forms of interphase area and flow regimes, great influence of small quantities of admixtures and small changes of geometry.

There are several methods to analyse diphasic flows.

The first approach is descriptive and experimental. It rests on empirical correlation. The method of correlation is the simplest form of analysis where regularities in behavior of experimental data are conveyed numerically, mostly with the help of dimensionless, physically valid complexes. However derivable regularities are applicable only for a very narrow scope of changeable parameters.

One of theoretical methods is a model of homogeneous flow. The real diphasic flow is a flow divided into areas with essentially different characteristics. Homogeneous theory of diphasic flows is a kind of try to present a flow as a monophasic one [1]. Such approximation begets mistakes that are corrected by involving supplementary members or coefficients in calculation equations.

In the model of separated flow [2] (sometimes it is called a model or theory of heterogeneous flow) each phase has its own characteristics (temperature, density, velocity) and should correspond to some form of normal laws of conservation of mass, impulse, energy.

The most well-grounded approach to the analysis of diphasic flow includes consideration of each phase as a continued medium and getting detailed description of three-dimensional field of flow. It may rest, for example, on use of two systems of vector equations of continuity and quantity of movement (for example, Navier-Stokes equations) together with boundary conditions and conditions of interphase area including mass carry effect and surface tension. Even if the form of all interphase areas is known (which usually never happens), in all cases except for few special cases, the solution of the problem takes great efforts.

In derivation of conservation equations for a certain scope there is freedom of choice to define some terms as basic and other terms as empirical. Particularly it's true for interphase interaction. The choice of equations form may influence the nature of solution. With all this going on the equations form changes with changes of the flow regime. And in order to take

IOP Conf. Series: Materials Science and Engineering **255** (2017) 012013 doi:10.1088/1757-899X/255/1/012013

into consideration all regimes of the flow and correction factors one needs a good deal of empiric information. The only criteria is utility for that very purpose which for all the analysis is carried out. All the alternatives need experimental check. Simple equations are unlikely possible and concern with only much idealized models which are geometrically simple and at best work as approximation for far more complicated real-world example.

The flow in the chamber of the phase separator is conditionally divided into diphasic core and turbulent layer which is δ thick. Friction in the interface is the source of resistance head of the flow.

The interface developed in the swirled flow on the curvilinear surface has a row of peculiarities in comparison with the flat flow. As the result of having longitudinal curve centrifugal forces spring up and hence the pressure gradient does in thickness of the interface. The picture of the flow in this case is similar in many aspects with the flow between rotating cylinders. In this case velocity distribution U is realized according to the law UR = const (the law of free swirl).

The system of differential equations of the spatial interface in incompressible liquid on the curvilinear wall generally in the presence of longitudinal and transversal gradients of pressure gives the momentum equation in projections on cylindrical coordinates:

$$\frac{\partial \delta_{\alpha R}^{**}}{\partial R} = \frac{\tau_{0\alpha}}{\rho U^2},\ \frac{\partial \delta_R^{**}}{\partial R} + \frac{1}{R}\left(\delta_R^* + \delta_\alpha^{**} + \delta_\alpha^* - \delta\right) = -\frac{\tau_{0R}}{\rho U^2}, \tag{1}$$

where R is radius of curvature, τ is friction tension, ρ is density, δ_R^{**} is momentum thickness of radial flow, δ_α^{**} is momentum thickness of circumferential flow , $\delta_{\alpha R}^{**}$ is momentum thickness of circumferential flow in the radial direction; δ^* is displacement thickness, δ is interface thickness.

The system (1) gave the system of quasilinear differential equations for partial derivatives of first order with two unknown quantities: δ_α^{**} and $\varepsilon = tg\theta_0$, where θ_0 is taper angle of ground current line.

The results of tests for visualization of ground current lines show that the taper angle of ground current line never depends on the radius and is practically equal to design value

$$\theta_0 \approx 59{,}4^\circ \left(tg\theta_0 = \varepsilon = 1{,}689\right).$$

Methods from theory of systems of quasilinear differential equations made it possible to carry out integration of equations for the case of circular current lines and of rotation according to the law of free swirl (C = UR). It gave the expression of momentum thickness in circumferential and core directions:

$$\delta_\alpha^{**} = 0{,}0786 \cdot \left(\frac{\nu}{U}\right)^{0{,}2} R^{0{,}8},\ \delta_z^{**} = 0{,}036 \cdot \left(\frac{\nu}{V_z}\right)^{0{,}2} L_K^{0{,}8} \tag{2}$$

where ν is kinematic density of the liquid, V_z is axial velocity, L_k is current length of the chamber.

These expressions make it possible to evaluate momentum thickness and therefore friction tension on the cylindrical wall of the tube.

Researching hydrodynamics of swirled flows from the point of view of mechanics of heterogenic mediums is a great problem that concerns first of all with the difficulties of solution of the ultimate task to define velocity and pressure fields. Such a problem can be solved only provided there are simplifications in balance equations of mass and momentum in the flows.

The course of transmitting air-enriched water through the phase separator gave separation gas-liquid mixture that could be rendered as a well-marked border of the phases. That's why the given mathematical model considers performance of two different phases.

The radial movement of the liquid in the rotating flow is ridden by centrifugal force, resistance force and also by some influence caused by incidental interaction of the particles and the flow.

The free surface of the phases is the surface of equal pressure. This surface is defined by character of pressure distribution in the liquid ring round radius and along the length and also by pressure distribution of in the gas vortex. As a rule the pressure change is ignored with minor velocity. The pressure is considered to be constant in this case. The border of phases separation gets fixed at the radius when pressure of the gas vortex and the liquid ring are equal.

To design a mathematical model of the swirled flow in the chamber of the phase separator the differential equations system of the movement of viscous incompressible liquid is taken as the original one.

The border of phases gets fixed at the radius when static pressure of the gas vortex and the liquid ring are equal.

The task to integrate equations of the movement of the swirled diphasic flow is considered in the presence of obligatory initial conditions: those of static pressure at the radius R_k; of total pressure at the radius R_k; of static pressure of gas in zero section. The radius of the chamber R_k should be established, mass gas consumption considered to be established and constant.

Generally the task of phases interaction in the swirled flow is characterized by seven equations

The block of equations to characterize the movement of the gas vortex consists of three equations:

- state equation
- energy equation of the gas flow (only axial velocity is taken into consideration)
- continuity equation for the gas flow

These equations define pressure, density and velocity in any section of the gas flow if the accepted constants of the total pressure, its mass consumption and the square of gas flow are known.

The formal parameter by means of which the liquid ring influences the course of the gas flow is the square of the open flow area of the gas vortex F_{gas} in the equation of the gas balance:

$$\frac{F_{gas}}{F_k} = \frac{P_k^{dyn}}{P_k^{total} - P_{gas}^{st}} \qquad (3)$$

where F_κ (F_r) is the square of the liquid ring, P_κ^{dyn} is the liquid dynamic pressure in the circumferential areas of the chamber, P_κ^{total} is the total pressure of the liquid, P_{gas}^{st} is the static pressure of the gas vortex.

IOP Conf. Series: Materials Science and Engineering **255** (2017) 012013 doi:10.1088/1757-899X/255/1/012013

What is evident from the analysis of the form (3) is that the higher the dynamic pressure in the circumferential area of the chamber is and the higher the static pressure in the gas vortex, the more the square of open flow the area of the gas vortex F_{gas} is.

The liquid performance is characterized by the following equations:

- energy equation
- continuity equation for the liquid ring
- Bernoulli differential equation (takes account of the axial constituent of the flow velocity in the liquid ring)

The system of these seven equations is closed and defines parameters of the gas-liquid flow in any section if the dynamic pressure of the liquid in the circumferential direction is known.

In order to turn from the section being considered to the nest section when doing integration, it is necessary to define changes of the dynamic pressure at the step Obviously it will decrease as the swirled flow under the influence of the viscosity will start losing its spin.

The dynamic pressure loss is explained by friction in the interface. The main characteristic of the interface is momentum thickness that depends upon the total velocity. It is necessary for the complicated three-dimensional flow to define the total momentum thickness δ_Σ^{**}, provided the sum of quantities for circumferential and consumed movements is equal to the quantity of the summarized movement. Having defined the value of δ_Σ^{**} with its components according to the equations (2):

$$\delta_\Sigma^{**} = \delta_\alpha^{**} \frac{U^2}{V_z^2 + U^2} + \delta_z^{**} \frac{V_z^2}{V_z^2 + U^2} \tag{4}$$

one gets tensions in the form of [3]:

$$\tau_\alpha = 0{,}01256 \left(\frac{U\delta_\Sigma^{**}}{\nu} \right)^{-0{,}25} \cdot \rho U^2,\ \tau_z = 0{,}01256 \left(\frac{V_z\delta_\Sigma^{**}}{\nu} \right)^{-0{,}25} \cdot \rho V_z^2, \tag{5}$$

where τ_α is friction tension on the wall in the circumferential direction, τ_z is in the axial direction.

Then the pressure loss is :

$$\Delta P_\kappa^{dyn} = \frac{(U\tau_\alpha + U\tau_z) 2\pi\rho\Delta l R_\kappa}{\dot{m}_{liq}} \tag{6}$$

Theoretically integration of parameters of the gas vortex can be derived to the endless length, provided its pressure can go down to zero endlessly and the velocity can increase endlessly. In fact the length of the chamber is limited and the hydraulic route behind the chamber of the phase separator doesn't allow pressure to go lower than the defined value that provides the established gas consumption. Generally in doing calculations for parameters of the flow in the chamber of the definite length L_κ with established initial parameters in the zero section the demanded pressure may be attained along the length which is less than the length of the chamber. That means that theoretically the gas vortex squeezed by the liquid ring must get destroyed. If this length approaches to zero, then the phase separating flow can't exist even in the zero section [4].

Pilot researches are the main test of the theory validity that allows applying the results of theoretical designs for practical purposes. In order to run pilot researches there was produced a pneudraulic test bench with multi-channel method of recording and processing measuring results on the basis of HS A/D transformer that makes it possible to run researches for hydrodynamics of the swirled flow in the chamber of the passive phase separator.

The test bench provides supply of the operating liquid (water) with various consumption (up to 0.65 kg/sec) and of the gas (air) with consumption up to $1.578 \cdot 10^{-3}$ kg/sec. The consumption of the operating liquid was controlled by the turbine flow sensor, the gas consumption was controlled by the rotameter. The water and the gas came through tubing to the mixer where the diphasic flow sprang up.

Pilot researches were divided into two stages. At the first stage we used the swirling chamber to measure the energetic parameters of the swirled liquid flow. This step was necessary for test a most important parameter of the algorithm, that determines the circumferential dynamic component of hydraulic head along the length of the swirling chamber under the influence of the viscosity in the boundary layer on the wall.

Then to visualize the diphasic flow and define main dependences between geometric and regime parameters the phase separator was used.

The research of the phase-separated gas-liquid flow was carried out in a special device that immediately visualizes the flow and measures main parameters at the inlet and the outlet.

The chamber of the phase separator was presented by a transparent tube with the internal diameter of 44 mm and the wall thickness of 13 mm. In the reservoir there was a tangential inlet of 8 mm to provide supply and spin of the operating body.

Table 1. The results of measuring total and static pressure of the liquid and the gas at the inlet and outlet of the plant with different consumptions of liquid and gas.

	Consumption kg/sec	$P^{st}_{liq\ in}$ $\cdot 10^5$, Pa	$P^{st}_{gas\ in}$ $\cdot 10^5$, Pa	$P^{st}_{gas\ out}$ $\cdot 10^5$, Pa	$P^{st}_{liq\ in}$ $\cdot 10^5$, Pa	D_{gas}, m	L_2, m
$m_l = 0,51$ kg/sec = const	$\dot{m}_{gas} = 1,578 \cdot 10^{-3}$	2,08	1,94	2,02	1,05	0,044 ÷ 0,028	–
	$\dot{m}_{gas} = 1,311 \cdot 10^{-3}$	1,92	1,80	1,92	1,02	0,044 ÷ 0,022	–
	$\dot{m}_{gas} = 6,939 \cdot 10^{-4}$	1,67	1,60	1,49	0,81	0,042 ÷ 0,018	–
$m_g = 3,44 \cdot 10^{-4}$ kg/sec = const	$\dot{m}_{liq} = 0,2815$	0,96	0,95	0,97	0,71	–	0,54
	$\dot{m}_{liq} = 0,2153$	0,61	0,62	0,67	0,48	–	0,394
	$\dot{m}_{liq} = 0,0978$	0,25	0,13	0,19	0,13	–	0,134

$P^{st}_{liq\ in}$ is statistic pressure of the liquid at the inlet, $P^{st}_{gas\ in}$ is statistic pressure of the gas at the inlet, $P^{st}_{gas\ out}$ is statistic pressure of the gas at the outlet, $P^{st}_{liq\ in}$ statistic pressure of the liquid at the outlet, D_{gas} is the diameter changes of the gas vortex by the length of the chamber, L_2 is the length of stable being of the gas vortex.

To research the character of the changes of the gas vortex diameter along the length of the chamber the consumption of gas was changed in the range of $3.45 \cdot 10^{-4} \ldots 1.58 \cdot 10^{-3}$ kg/sec with the constant consumption of the liquid which is equal to $\dot{m}_l$ = 0.51 kg/sec. For each gas consumption one measured total and static liquid pressure in the circumferential area (on the wall) in the initial section of the phase separation (see Table 1) and also photographed to measure the diameter of the gas vortex (Fig. 2).

Then to define the dependence of length of the gas vortex (or length of the diphasic swirled flow being) on consumption of liquid with the constant gas consumption $\dot{m}_g = 3{,}44 \cdot 10^{-4}$ kg/sec one changes liquid consumption in the range of 0,098…0,282 kg/sec. With the help of photos the length of the gas vortex was defined (Fig. 3).

The analysis of calculation and experimental data showed that accuracy of calculation algorithm is satisfactory and is not higher than 5% in comparison with experimental results.

The results of calculations and experiments gave dependence (Fig. 4) of the gas vortex length (or the length of the diphasic swirled flow being) on the liquid consumption that physically corresponds to the circumferential velocity of liquid in the initial section.

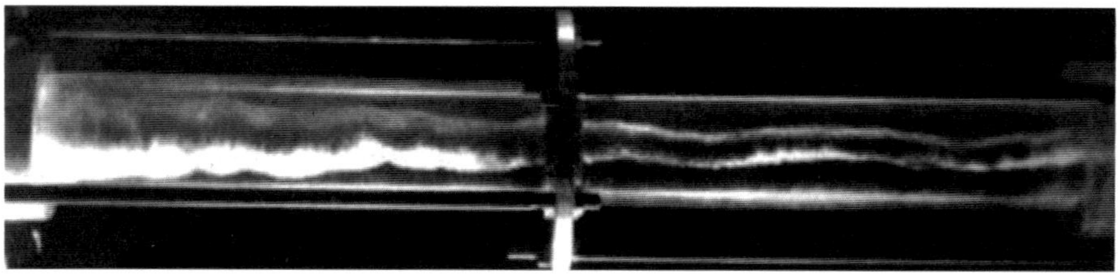

Figure 2. The swirled diphasic flow with gas consumption $\dot{m}_{gas} = 3{,}45 \cdot 10^{-4} kg/\sec$. Changes of the gas vortex length 0,042…0,016 m

The gas content in the flow constitutes 0,07 %.

Pilot and calculation researches show:

- the gas vortex length in the diphasic flow with the constant gas content decreases along the length of the tube as the result of liquid spin decrease caused by breaking effect of the wall;
- the length of the swirled diphasic flow being in the round tube definitely depends on the flow spin in the initial section, the spin decrease leads to decrease of the length;
- consumption decrease leads to the fall of axial velocity and axial velocity of liquid remains almost unchanged;
- changes of gas static pressure happen only in the initial area, and then remains constant along the length of the chamber.

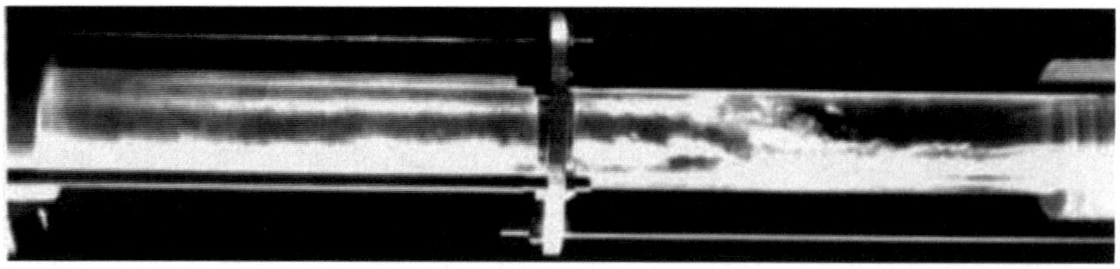

Figure 3. The length of the stable gas vortex $l_{gas.v.}$= 0.394 m. Liquid consumption $\dot{m}_{liq} = 0{,}215 kg/\sec$. The gas content constitutes 0.16 %

The designed methods allow to evaluate interdependence of the basic parameters of the swirled monophasic and diphasic flows with tangential supply of liquid and gas-liquid mixture and to define influence of geometry of the chamber flowing part on the basic parameters of the flow.

It is significant that the offered algorithm of calculation for the flow in the chamber of the phase separator takes into account changes of liquid characteristics along the length of the chamber caused by temperature changes of the operating liquid.

The developed model lets not only calculate the nominal regime of the phase separator functioning but also helps foresee possible breakdowns and in advance take measures to eliminate the revealed defects in the process of its designing.

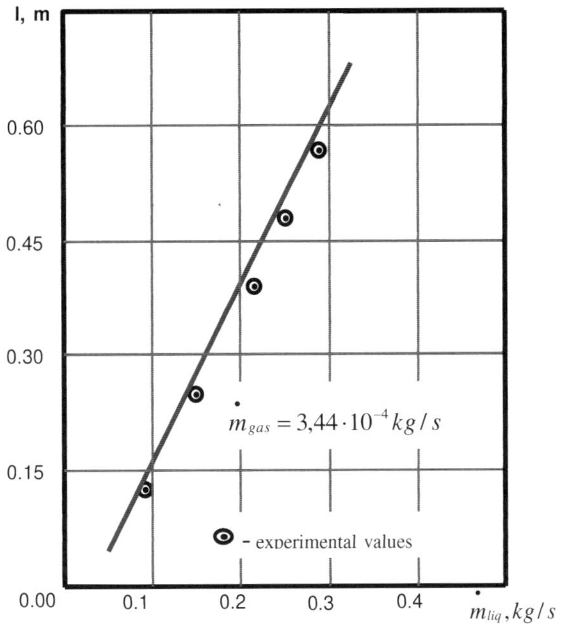

Figure 4. The dependence of the gas vortex length on liquid consumption

Further research of flow regularities for heat exchange and mass transfer of swirled flows in axial-symmetric channels, systematization of these data and designing universal calculation methods for such flows are topical scientific and practical problem. This research results will be widely applied for purposes of various engineering fields.

References

[1] Chisholm D 1983 Two-phase flow in pipelines and heat exchangers (London, New York) p 304
[2] Wallis G B 1969 One Dimensional Two-phase Flow (McGraw Hill, New York) p 408
[3] Kraev M V, Kishkin A A, Melkozerov M G 2002 Hydrodynamics of a Two-phase Flow in Phase Separator Chamber. Russian Aeronautics (Allerton Press, Inc./New York) vol **45** No 3 pp 30-37
[4] Melkozerov M G, Delkov A V, Chernenko E V, Litovchenko V I 2010 Modeling for Swirling Diphasic Flow in the Centrifugal Phase Separator. Mathematical Modeling and Computer Simulation of Material Technologies *Proceeding of the 6 International Conference MMT Ed. By M. Zinigrad* (Ariel University Center of Samaria, Israel) pp 1-75 – 1-84

Formation of dislocation loops and voids in electron irradiated zinc selenide single crystals

Y Y Loginov[1], P D Brown[2], I V Kovalev[1]
[1] Reshetnev Siberian State University of Science and Technology, 31 Krasnoyarskiy Rabochiy av., Krasnoyarsk, 660037, Russia
[2] University of Nottingham, Nottingham, NG7 2RD, UK

E-mail: loginov@sibsau.ru

Abstract. The formation of small dislocation loops in size of 2.5 - 45 nm and a density of $1.4 \cdot 10^{11}$ cm^{-2}, as well as voids in size ≤10 nm in ZnS crystals were found by the transmission electron microscopy method (TEM). Samples were examined and irradiated *in situ* in a JEOL 4000EX-II electron microscope operated at energy of 400 keV and intensity of (1 - 4)$\cdot 10^{19}$ e/cm$^2 \cdot$s. Fine particles of a new phase in size ≤10 nm are formed also. These features can be identified from an analysis of moiré fringe contrast as phase of ZnO_2. Similar defects in single crystals of ZnS formed *in situ* after irradiation in a transmission electron microscope JEM-100CX operated at energy of 100 keV and intensity of $3{,}5 \cdot 10^{17}$ e/cm$^2 \cdot$s. It was found that the formation of structural defects in ZnS under electron irradiation with above-threshold energy of defect formation (400 keV) is similar to the formation of structural defects in these crystals under electron irradiation with sub-threshold defect formation energy (100 keV).

1. Introduction

The study of the formation of structural defects and changes in the properties of irradiated semiconductors is of interest both in terms of increasing their radiation resistance, and ability to manage the properties of materials under irradiation [1-4]. This is especially important for semiconductor compounds A_2B_6, because of their high sensitivity to radiation and insufficient study of these issues.

Zinc sulfide, belonging to the group of compounds A_2B_6, is used for the detection of charged particles. The high efficiency of radiative recombination of electrons and holes of the material makes it possible to use it as an active element of the semiconductor laser. Further, ZnS is a promising material of the power quantum optics.

The formation of structural defects in semiconductors during electron irradiation is possible in the conditions of electron irradiation with a high intensity ~10^{17}-10^{19} e/cm$^2 \cdot$s that is achievable, for example, in the transmission electron microscope (TEM). These intensities several orders of magnitude higher than the intensity of radiation in conventional industrial accelerators (~10^{12}-10^{14} e/cm$^2 \cdot$s), which means that in TEM on the area occupied by one atom drops ~ 10^2 - 10^4 electrons per second.

The formation of structural defects in materials irradiated by electrons with energies above the threshold energy of radiation-induced defects formation (E_T) under irradiation *in situ* in a high-voltage electron microscope ($E \geq 1$ MeV) has been well studied for Si, Ge [3, 5] and metals [6, 7]. The formation of structural defects in ZnS under electron irradiation with above-threshold energies ($E > E_T$) has been insufficiently studied.

Influence of electrons with sub-threshold energies ($E < E_T$) on the processes of structural transformations in semiconductors has been studied in [8-11] where it was reported about possibility of formation of point defects (vacancies and interstitials) in these materials under such irradiation. The formation of structural defects under irradiation of ZnS crystals by electrons with energy $E < E_T$ in TEM previously discussed in [12]. In the interpretation of the results was used idea about the displacement of the crystal lattice ions under the influence of electrons with sub-threshold energy as a result of the ionization mechanisms of point defects formation based on the known mechanism of Varley [8].

The purpose of this work was to study the influence of electrons with an energy of 400 keV ($E > E_T$) and 100 keV ($E < E_T$) on the formation of structural defects in ZnS.

2. Experimental

Crystals of ZnS with resistivity $\rho \approx 7 \cdot 10^3$ Ω·cm were prepared using a modified Piper-Polich technique. The crystals were synthesized directly from the vapour phase at a temperature of 1500 ^{0}C using polycrystalline starting material, under an argon ambient. Bulk ZnS crystals were prepared in thin foil form by sequential mechanical polishing and argon ion milling at 5kV, 20μA and 15^0 with liquid nitrogen cooling until electron transparent. Improved sample foils were obtained following iodine reactive ion sputtering for 5 to 10 min at 3kV, 10μA and 15^0 at room temperature. To completely remove the surface defects formed by ion milling as appropriate thin foil was subjected to chemical polishing for 2-3 seconds in a dilute solution of HPC (a supersaturated solution of CrO_3 in H_3PO_4 heated to 60 ^{0}C (1 part), mixed with concentrated HCl (2 parts)). Samples were examined and irradiated in a JEOL 4000EX-II electron microscope operated at 400keV ($E > E_T$) and $j = (1 - 4) \cdot 10^{19}$ e/cm^{2}s and JEM-100CX electron microscope operated at 100keV ($E < E_T$).

3. Results and Discussion

The electron microscopic study of ZnS crystals before irradiation show that they contain of well developed native dislocations and long stacking faults. Also well-developed parallel strips (striations), indicating polytypic modifications of ZnS, are also observed in all cases. This is especially true when the diffraction vector **g** is perpendicular to these strips.

Irradiation of ZnS by electrons with an energy of 400 eV in TEM accompanied by the formation of small dislocation loops. Figure 1,a shows the formation of dislocation loops in ZnS after electron irradiation for 5 minutes. Dislocation loops have a size of 2.5 - 45 nm at a density of $1.4 \cdot 10^{11}$ cm^{-2}. In addition, striations due to the polytype structure of this compound are more readily apparent. The ZnS foil shown in Fig.1 was given a brief chemical polish following Ar^+ and I^+ milling to remove the remnant artefact structure.

Subsequent electron beam irradiation of ZnS caused growth, transformation, shrinkage and movement of dislocation loops, the formation of new loops, and then the formation of voids. Figure 1b shows the bright-field image of ZnS after 15 minutes of irradiation electrons. It is seen the formation of fine voids in the form of bright points of 3-8 nm in size.

Figure 2,a shows a high resolution image of voids formed in ZnS after irradiation for 15 min. It is seen the formation of voids, which are formed by the accumulation of vacancies, indicating that further decomposition of the material under electron irradiation. At the same time, the crystal structure of the material is preserved, as indicated by electron diffraction pattern obtained from the same place (Fig. 2b).

Formations of fine particles size ≤ 10 nm with specific moiré fringe contrast were also observed. Creation of modern semiconductor devices of ultra-sizes requires control not only

IOP Conf. Series: Materials Science and Engineering **255** (2017) 012014 doi:10.1088/1757-899X/255/1/012014

the presence of defects such as dislocations, dislocation loops, stacking faults, and microtwins, but also control defects of size of ≤ 10 nm. As shown earlier, the formation of new phases in such sizes in materials A_2B_6 possibly under the influence of electrons and ions [12, 13].

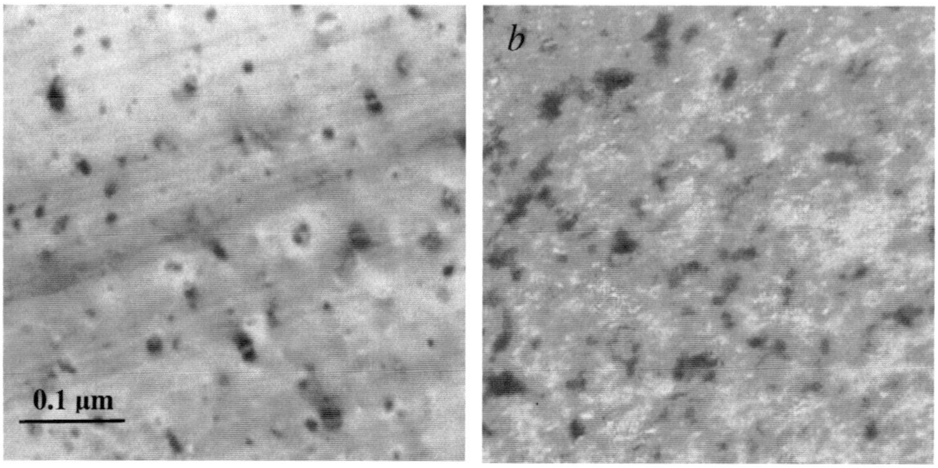

Figure 1. TEM image of structural defects in the ZnS irradiated with 400 keV electron beam of intensity $1 \cdot 10^{19}$ e/cm^{2}s for 5 minutes (a) and 15 minutes (b)

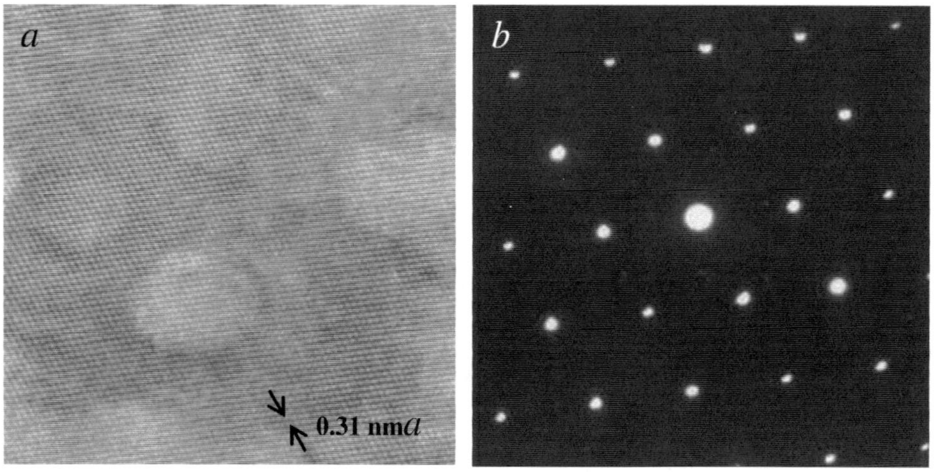

Figure 2. <110> HREM image of 400 keV electron beam irradiated ZnS with intensity $4 \cdot 10^{19}$ e/cm^{2}s for 15 minutes (a). Diffraction pattern from the same place of thin foil (b)

Typically, microanalysis of the particles of this size is difficult to technically implement. However, the use of technique moiré contrast, formed from the particles in the high-resolution images, allowing, in certain cases, to determine the chemical composition of these particles. In this case, the period of the moiré contrast D, occurring when rotating the lattices with the parameters d_1 and d_2 are at an angle α, can be determined according to [14]. As a result of analysis of fine particles formed in ZnS crystals, irradiated by electrons with energy of 400 keV, was found that the small particles with a moiré fringe contrast can be identified as the phase of ZnO_2, which is consistent with previous results [1]. So, the decomposition products produced by 400keV electron beam irradiated ZnS is a ZnO_2. The tendency for the initial formation of oxides of Zn indicates the preferential removal of the anion leading to accelerated oxidation [15].

Similarly to ZnS irradiated by electrons with energy $E > E_T$, the irradiation of these crystals by electrons with energies $E < E_T$ is also accompanied by the formation of dislocation loops and voids.

In Fig.3 shows the formation of structural defects in ZnS irradiated by electron beam with energy of 100 keV for 10 minutes. Defects in the form of small clusters with dark contrast and the dislocation loops are generated first in the center of the irradiated area, and then spread to the periphery. The density of defects increases with the increase of the exposure time. Along with these defects are formed voids with sizes from 3 to 8 nm in diameter, which is clearly seen in Fig.3 b at higher magnification.

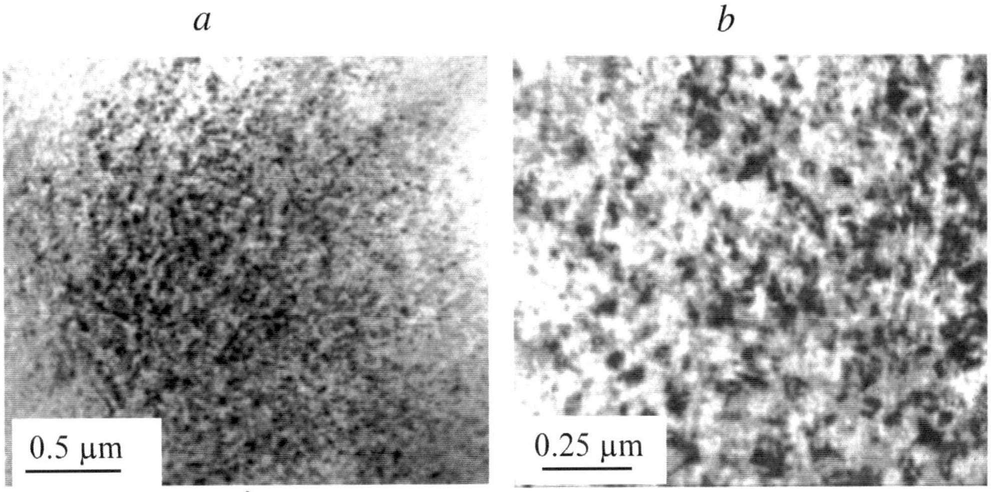

Figure 3. The formation of structural defects in ZnS irradiated with electrons with energy of 100 keV and the intensity of $3\cdot10^{17}$ e/cm^{2}s for 10 min.

The final form, concentration a d distribution of defects depend on the dose of electron irradiation, the sample temperature and the elastic stress fields. The rate of defect formation is directly proportional to the electron beam intensity, while the final form and density of the secondary defect does not depend on the intensity of exposure. The same patterns are observed in ZnS irradiated by electrons with energies above the threshold energy of radiation-induced defects formation (400 keV).

The experimental data are well explained in the framework of sub-threshold ionization mechanism of defect generation and the subsequent interaction of point defects. Multiple anion ionization causes the formation of Frenkel pairs: interstitial atoms and vacancies. Moderate heating of the sample by the electron beam and the effect of elastic stress fields determine the rate and distribution of secondary structural defects in the irradiated material. Different mobility of interstitial atoms and vacancies creates the possibility of their spatial separation and the formation of vacancy and interstitial clusters. Formation of clusters of point defects in crystals under electron irradiation, in particular, dislocation loops, due to separation of the components of Frenkel pairs due to different mobilities of vacancies and interstitials.

Voids formation in chalcogenides reflects the fundamental processes of decomposition occurring in the material that is apparently originally associated with atoms of VI group (S) in the formation of interstitial-type clusters with subsequent preferential removal chalcogens from these clusters, as they may exist in the gaseous state and the heat of evaporation is less than the heat of vaporization of Zn atoms. As a result, their place will remain the voids, which will further decorated with a metal phase in the process of migration of Zn atoms on these sinks (or, if the Zn atoms were part of this cluster, they will be concentrated within the void).

4. Conclusion

It was found that the irradiation of ZnS crystals by electrons with energy of 400 keV is accompanied by the formation of dislocation loops, and further shrinkage and movement of loops, the formation of voids and fine particles of the new phase, which can be identified from an analysis of a moiré fringe contrast as ZnO_2.

The phenomenon of the dislocation loops and voids formation in the ZnS crystals under sub-threshold electron irradiation is observed also. The beginning of the formation discernible in the TEM structural defects, the types and concentrations are a function of dose and temperature of electron irradiation and depend on the fields of elastic stresses in the crystal. The experimental data are well explained in the framework of sub-threshold ionization mechanism of defect generation and the subsequent interaction of point defects. A pattern of defects similar patterns of defect formation in these crystals irradiated with electrons with sub-threshold energies. It was found that in the defect formation in the crystals of ZnS irradiated by electrons with energies above threshold along with elastic collisions is significant contributions and inelastic collisions.

Regularities of structural defects formation in ZnS can be used to solve problems of management type, density and spatial distribution of defects in the crystal structure, which is important for the implementation of the limiting parameters of microelectronic devices.

References

[1] Loginov Y, Brown P, Durose K 2003*The Regularities of Structural Defect Formation in Semiconductors II–VI* (Logos, Moscow)
[2] Kozlovskii V, Lebedev A, Bogdanova E, Seredova N 2015 *Phys. Techn. Semicond.* **49** 1198
[3] Aseev A, Fedina L, Hoehl D, Bartsch H 1994*Clusters of interstitial atoms in silicon and germanium* (Akad. Verlag, Berli,)
[4] Grushka O, Maslyuk V, Chupyra S, Mysliuk O 2012 *Phys. Techn. Semicond.*, **46** 327
[5] Hirata M and Kiritani M 1983 *Phys.B.* **116** 616
[6] Nakai K, Kinoshita C, Muroo Y, Katajima S 1983 *Phil. Mag. A.*, **48** 215
[7] Urban K 1979 *Phys. Stat. Sol.(a)*, **56** 157
[8] Varley J 1954 *Nature*, **174** 886
[9] Vavilov V, Kiv A, Niyazova O 1975 *Phys. Stat. Sol.(a)*, **32** 11
[10] Yunusov M, Zaikovskaya M, Oksengendler B, Tokhirov K 1976 *Phys. Stat. Sol.(a)*, **35** K145
[11] Karpov V and Klinger M 1987 *Phys. Tech. Sem.*, **12** 1887
[12] Loginov Yu and Brown P 1992 *Phys. Stat. Sol. (a)*, **132** 323
[13] Loginov Y, Brown P, Humphreys C 1995 *Microsc. Semicond. Mater. Bristol, London: Inst. Phys. Conf.*, **146** 431
[14] Fultz B and Howe J 2011 Transmission electron microscopy and diffractometry of materials (Tehnosphera, Moscow)
[15] Thangaraj N and Wessels B 1990 *J. Appl. Phys.*, **67** 1535

Lorentz microscopy methods for magnetic domain structure study

M N Volochaev[1, 2], L I Kveglis[3], Yu Yu Loginov[2]

[1]Institute of Physics, Russian Academy of Sciences, Siberian Branch, Krasnoyarsk 660036, Russia
[2]Reshetnev Siberian State University of Science and Technology, Krasnoyarsk 660014, Russia
[3]Siberian Federal University, Krasnoyarsk 660041, Russia

E-mail: volochaev91@mail.ru

Abstract. The capabilities of transmission electron microscope Hitachi HT7700 for magnetic domain structure (MDS) study of thin films using Lorentz transmission electron microscopy (LTEM) methods is discussed. Two methods of magnetic domain structure study (defocus method and aperture shift method, including low-angle electron diffraction (LAED)) in thin magnetic films were examined.

1. Introduction

Modern methods of transmission electron microscopy (TEM) can solve quite a wide range of research tasks in the field of solid state physics and materials science. Around TEM capabilities LTEM occupy a special place. Lorentz microscopy methods allow to obtain a direct MDS images of thin films magnetic materials. Nowadays, a number of researches describe both MDS study [1-6] and methodical reports on electron microscopes adaptation for LTEM [7-9]. Using specifically designed sample holders allows to *in situ* study of MDS evolution depending on temperature [1-3], external magnetic field [3-5], strain etc. The defocus method is usually used to MDS study, in which domain walls are observed as light and dark contrast on magnetic domains boundaries [10]. Nevertheless, there is the so-called 'aperture shift method'[10] (including LAED method), in which magnetic domains are directly observed instead the domain walls. The aperture shift method is the following: the electron beam transmission through the magnetized sample along the ***y*** axis, the electron beam is deflected by the Lorentz force along the ***x*** axis. The angle of electron beam deflection by the sample magnetic field is determined by the expression:

$$\varphi_x = \frac{4\pi\tau}{c}\left(\frac{e}{2Um}\right)^{\frac{1}{2}} M_y \tag{1}$$

where φ_x is the angle of electron beam deflection by Lorentz force along the ***x*** axis, τ is sample thickness , e is electron charge, m is electron mass, U is acceleration voltage along the z axis and M_y is the sample magnetization projection along the ***y*** axis.

Since the electron beam deflection has different directions in the adjacent magnetic domains, transmitted through the magnetized sample electron beam is split into several parts.

This split can be directly observed on the LAED patterns for both direct and diffracted beams. Aperture can close one of the split beams. Alternate domains look like light and dark areas. The dark domains correspond to the beam which is closed by aperture. Both methods in detail and the theory of the magnetic contrast origin are described in [10,11].

In this report we describe capabilities of conventional TEM Hitachi HT7700 (acceleration voltage up to 120 kV) for Lorentz microscopy methods (defocusing and aperture shift).

2. Methods

MDS images and LAED patterns were obtained by only changing of microscope lens currents. The major requirement for performing LTEM is external magnetic fields absence in the sample region. Therefore, the first stage is to close the current supply to the objective lens, and use the objective mini-lens instead. The objective mini-lens magnetic field does not influence on the sample magnetization [7]. Incident electron beam must be as possible more parallel for achieving clear MDS images in LTEM mode, thus condenser lens system is defocused for efficiently producing a point source at a large distance [10].

The second stage is achieving maximum possible magnification by maximum increase the current of projection lenses while maintaining sufficient brightness to obtain high-quality MDS images. The optimal combination of "magnification degree - brightness" is achieved in 'LOW MAG' mode and consist about 2kX (at 100 kV acceleration voltage) for Hitachi HT7700. The domain wall width image is about 100 nm in these conditions. It is much higher than the known magneto-optical MDS study methods. Intermediate lenses are used to focus MDS image.

The third step is adjusting electron beam and sample image on the optical axis of the microscope by deflecting coils of intermediate lens and column. Intermediate lens deflecting coils shift the sample image in X-Y plane, objective mini-lens rotate image around z axial. This step is necessary to overlap LTEM and TEM images that to study correlations between domain walls and lattice defects location in thin magnetic foils or epitaxial films. As a result, magnetic contrast [10] can be observed with magnification from 100 to 2kX.

Observation LAED patterns also occurs when the current supply of the objective lens is closed. Splitting diffraction peaks (satellites) expedient to study on central undiffracted beam because it has the greatest intensity than diffracted beams intensity. Satellites will be observed only if the selected area contains at least two adjacent magnetic domains. Satellites are quite small (about 10^{-4} rad) depends mainly on the magnetization magnitude and film thickness (at the same accelerating voltage). Effective camera length must be increased as far as possible (about 20m) by adjusting projection lens that satellites are observed. Shifting the aperture can close one of two split beams. Then the alternate magnetic domains look like light and dark areas, at that, the dark domains correspond to the beam which is closed by aperture. Magnetic domain images thus are similar dark-field TEM images. Objective aperture used as selective aperture but selective aperture used to close one of split diffraction peaks.

The advantage of this method lies in obtaining images of magnetic domains and crystalline structure in a precise focus. The aperture shift method allows study as magnetic domains as focused lattice defects at the same time.

3. Materials

Fe and Ni magnetic thin films were chosen for Lorenz microscopy methods study. Starting materials (Fe and Ni) purity was better than 99.95%. Fe thin films were obtained by the thermal deposition at a of 300 °C onto NaCl (001) substrate in a vacuum at a residual pressure of 10^{-6} Torr. Ni thin films were obtained by magnetron sputtering at room temperature onto NaCl (001) substrate. The base chamber pressure and total sputtering pressure were 10^{-6} Torr

IOP Conf. Series: Materials Science and Engineering **255** (2017) 012015 doi:10.1088/1757-899X/255/1/012015

and 6×10^{-3} Torr respectively. Fe and Ni film thickness was about 30 nm. Then films were separated from the substrates in distilled water and deposited onto the supporting copper TEM grid. Magnetic domain structures were studied at 100 kV accelerating voltage.

4. Results and discussion

4.1. Defocus method

Fig. 1 shows MDS (domain walls) observation scheme by defocus method [10]. The electron beam is deflected at the opposite directions with divergence angle 2φ (see Eq. 1) in adjacent magnetic domains. Intensity excess is observed in region 1", intensity lack is observed in region 2", when sample image is observed in A" - B" plane (under-focus condition). Reverse contrast (region 1' is intensity lack, region 2' is intensity excess) is observed in A' - B' plane (over-focus condition). The contrast is alternating light and dark bands onto magnetic domain boundaries (so-called domain walls). Domain wall contrast disappears at in-focus condition (A - B plane). It should be noted that the width of the domain wall image is dependent not only on the defocus distance, but also on the film thickness and the magnetization distribution within the domain wall [10].

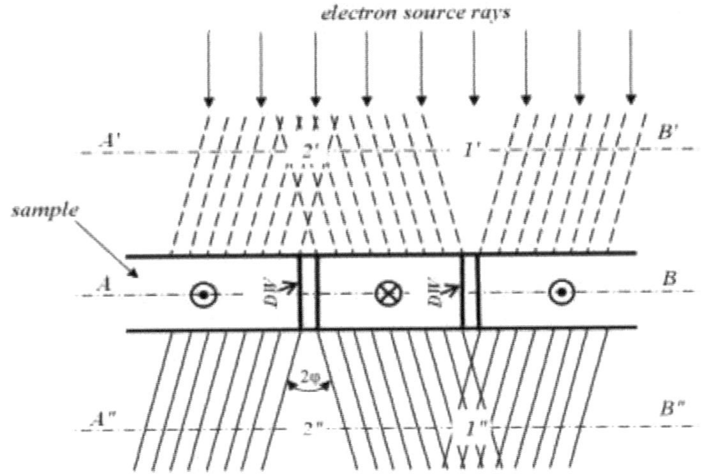

Figure 1. The Observation scheme of magnetic domains by defocus method (DW - domain wall)

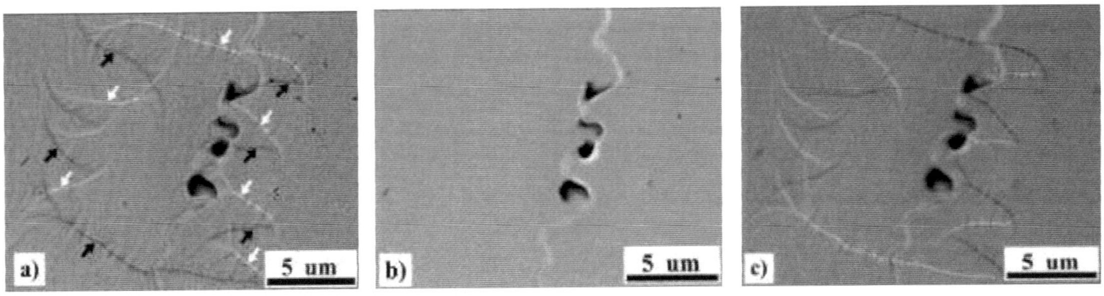

Figure 2. Magnetic domain images of Ni thin film was obtained at different observation conditions: a - over-focus, b - in-focus and c - under-focus.

Fig. 2 shows magnetic domain images of Ni thin film at over-focus (A' - B' plane (see fig. 1)), in-focus (A - B plane) and under-focus (A" - B" plane) conditions.

White and black arrows (see. Fig. 2a) show magnetic domain boundaries, where excess (collecting domain walls) and lack (divergent domain walls) of intensity are formed respectively. Figure 2b shows that magnetic contrast absent at in-focus condition. Figure 2c shows the same area image of Ni thin film with a reverse magnetic contrast (A" - B" plane fig. 1).

Fig. 3 shows classic example of vortex magnetic structure formation in Fe thin film upon objective lens external magnetic field application. External magnetic field is perpendicular to the film plane. The "fibration" or magnetization ripples is clearly visible besides the main domain walls. "Fibration" is characterized for magnetic films having magnetic anisotropy dispersion, including the normal direction to the film plane [12]. Red arrows indicate the magnetic poles - the areas in which the magnetization vectors perpendicular to the sample plane but opposite on direction.

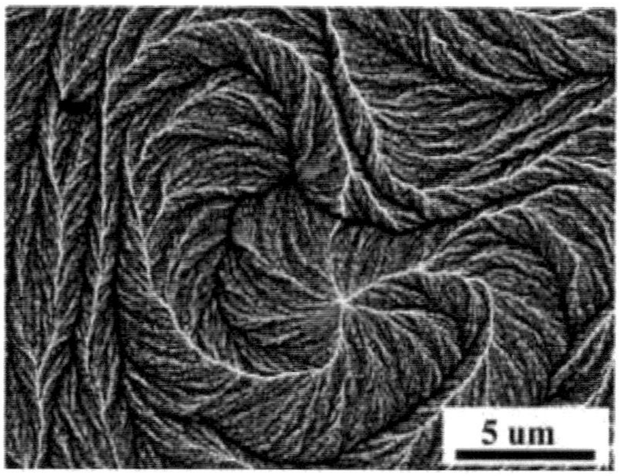

Figure 3. The vortex magnetic structure image of Fe thin film (under-focus condition).

The main advantage of defocus method is the capability to determine the local magnetization vector direction in the film plane by the "fibration" direction. However, this method does not allow to observe magnetic contrast and focused lattice defects image at the same time.

4.2. Aperture shift method

Fig. 4 shows MDS observation scheme by aperture shift method. Since current supply of main objective lens is switched off, diffraction image is formed in selective aperture region. Selective aperture used to isolate one of the split LAED peaks. This is analogically operation with objective lens aperture in TEM mode. The objective lens aperture used to select the sample region, which corresponds to the LAED pattern. Minimum aperture size of objective lens Hitachi HT7700 is 10 μm in diameter. Consequently, the sample area size corresponding LAED is about 10 μm.

Fig. 5a shows magnetic domain image of Fe thin film obtained by defocus method. LAED pattern (insert of Fig. 5a) allow to determine magnetization direction of adjacent magnetic domains, if zero diffraction peak corresponding un-split reflex (e.g. LAED pattern was obtained from blank area) was superposed with LAED pattern of adjacent magnetic domains.

Straight-line tension bar between double diffraction peaks mean that domain walls are Bloch walls. Tension bar have arcuate shape if domain walls are Néel walls.

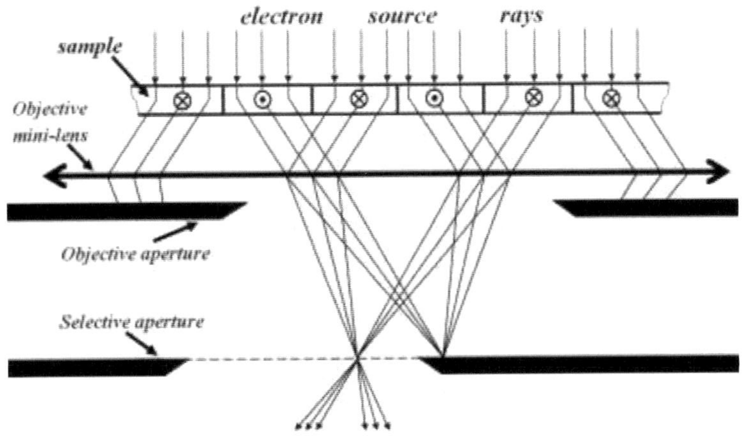

Figure 4. The Observation scheme of magnetic domains by aperture shift method.

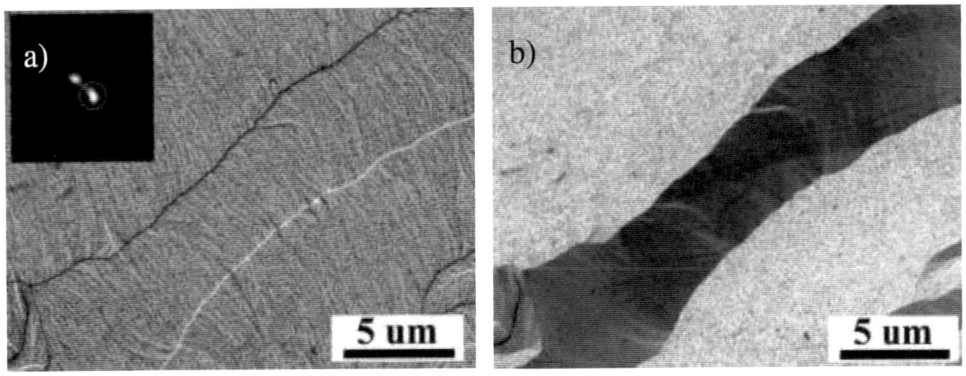

Figure 5. Magnetic domain images of Fe thin film was obtained by (a) defocus method (The insert shows the corresponding LAED pattern) and (b) aperture shift method.

However, Neel walls are formed at low film thickness (about 10 nm), therefore deviations are quite small to determine the domain wall type by the tension bar shape. For 180-degree magnetic domains value of local film magnetization possible to determine by distance between double diffraction peaks if the film thickness is known and vice versa [10]. Fig. 5b shows image of the same film area obtained by aperture shift method. Domain corresponding to the closed beam (circled in red on insert of Fig. 5a) has become dark. Magnetic inhomogeneity are visible inside dark domain. In addition there is an inversion in the phase banded contrast, the nature of which requires further study.

Fig. 6 shows magnetic domain images of one and the same area of Ni thin film. Fig. 6a is magnetic domain image obtained by defocus method, the inset shows the corresponding

LAED pattern. The distance between the split diffraction peaks of Ni thin film is 3.5 times less than Fe thin film which is consistent with the published data on the magnetization measurement for Fe and Ni [10, 11]. Fig. 6b and fig. 6c are magnetic domain images obtained by aperture shift method and correspond to the closed upper and lower diffraction peaks (see insert of fig. 6a), respectively.

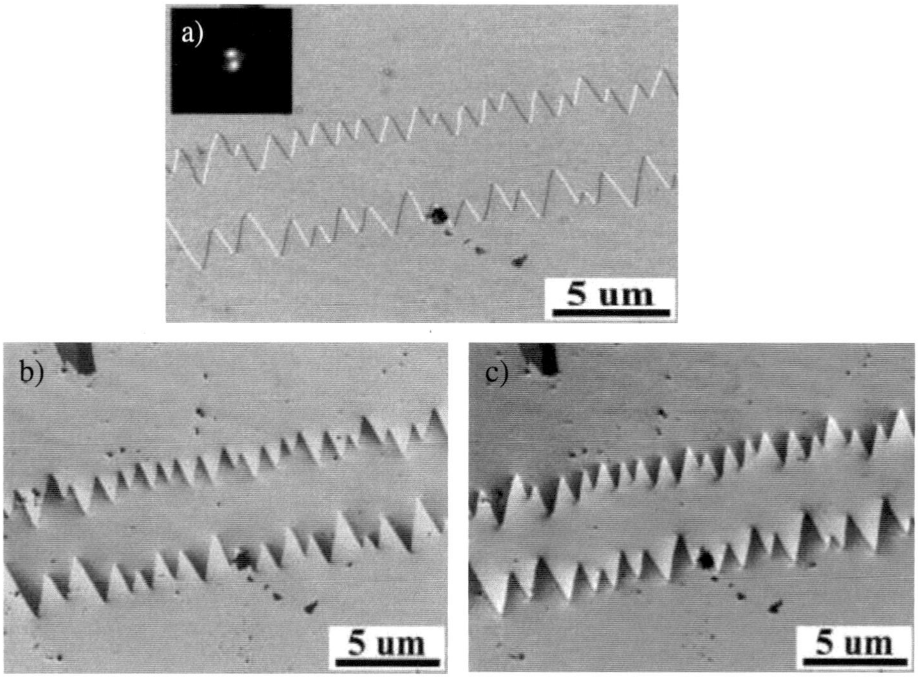

Figure 6. MDS images of Ni thin film obtained by (a) defocus method (the insert shows the corresponding LAED pattern) and (b, c) aperture shift method.

LAED method is most preferred to determine the magnetization vector direction in thin foils and epitaxial films (due to lack of "fibration" which characterized for polycrystalline and amorphous films).

5. Conclusion

We demonstrated capabilities of Hitachi HT7700 for MDS study of thin films. The combination of defocus, LAED and aperture shift methods can provide full information not only about the size and shape of the magnetic domains as well as the local distribution of the magnetization vector direction in the film plane. The features of LAED method can establish the correlation between domain walls and crystal lattice defects.

Acknowledgments

Electron microscopy studies were performed by the equipment of Center for Collective Use Krasnoyarsk Scientific Center, Russian Academy of Sciences, Siberian Branch. The authors would like to acknowledge Dr. Zhigalov V.S., Kirensky Institute of Physics for assistance in obtaining specimens for Lorentz electron microscopy studies.

References

[1] Saito K, Park H, Shindo D, Yoshizawa Y 2006 *Journal of Magnetism and Magnetic Materials*, **305** 304
[2] Bartova B, Wiese N, Schryvers D, Chapman J, Ignacova S 2008 *Acta Materialia*, **56** 4470
[3] Akase Z, Aizawa S, Shindo D, Sharma P, Makino A 2015 *Journal of Magnetism and Magnetic Materials*, **375** 10
[4] Kasahara T, Park H, Shindo D, Yoshikawa H, Sato T, Kondo K 2006 *Journal of Magnetism and Magnetic Materials*, **305** 165
[5] Budruk A, Phatak C, Petford-Long A, Graef M 2011 *Acta Materialia*, **59** 6646
[6] Kim J, Park H, Shindo D, Iseki T, Oshimura N, Ishikawa T, Ohmori K 2007 *Materials Transactions*, **48** 2606
[7] Kohn A, Habibi A 2012 *JEOL News*, **47** 17
[8] Schofield M, Beleggia M, Zhu Y, Pozzi G 2008 *Ultramicroscopy*, **108** 625
[9] Schofield M, Beleggia M, Lau J, Zhu Y 2007 *JEOL news*, **42** 2
[10] Hirsch P, Howie A, Nicholson R, Pashley D and Whelan M *Electron microscopy of thin crystals* (Mir, Moscow, 1967)
[11] Tomas G, Goringe M *Transmission electron microscopy of materials* (Wiley, New York 1990)
[12] Fuller H, Hale M 1960 *J. Appl. Phys.*, **31** 238

IOP Conf. Series: Materials Science and Engineering **255** (2017) 012016 doi:10.1088/1757-899X/255/1/012016

On the Static Accuracy of Charge-Discharge Units Intended for Electrical Tests of High Capacity Li-ion Batteries

E A Mizrah, D K Lobanov, E A Kopylov, R V Balakirev, A S Fedchenko
Reshetnev Siberian State University of Science and Technology
31, Krasnoyarskiy rabochiy ave., Krasnoyarsk, 660014, Russia

E-mail: enis-home@mail.ru

Abstract. Performing of the cycle testing according to the principles of Dynamic Stress Test can significantly reduce the overall time of development and production of batteries, which in turn allows reducing the cost of designing and testing of the spacecraft power systems. Performing of Dynamic Stress Test require special charge-discharge units that allows to perform a full cycle of electrical tests of batteries, including cyclic testing. Providing the required accuracy of measurement and stabilization of certain attributes of Li-ion battery operating modes is one of the problems that arise during the development of such charge-discharge units. The following attributes are of particular interest: charge and discharge currents, discharge powers, battery voltages. Analysis of the charge-discharge unit as a control system allows evaluating the steady-state stabilization error of the required attributes of the developed device. Moreover, using a digital integrator in the control system of the charge-discharge unit allows providing specified values of steady-state stabilization error of required attributes in different test modes.

1. Introduction

Li-ion batteries are widely used in spacecraft power systems. Characteristics of Li-ion batteries significantly affect the spacecraft life cycle in orbit. Electrical tests of Li-ion batteries are performed during development and production at the factory, in particular, the battery cycle life is estimated by special designed cycle tests. These tests include multiple charge/discharge cycles in order to reduce total capacity and power of the batteries to a minimum value.

Reducing the time of the cyclic tests can significantly speed up and reduce the cost of designing and testing of spacecraft power systems. In order to reduce the duration of cyclic tests were developed methods of cycle tests for Li-ion batteries based on the method of Dynamic Stress Test (DST) [1–3]. This method involves increasing of charge-discharge currents of the tested battery up to the maximum values, including modes of discharging by constant power. Charging of batteries is typically performed according to the manufacturer's methods, discharging - by the techniques of DST.

There are many types of charge-discharge units (CDU) for Li-ion batteries [4–8], but these devices are not suitable for testing (including cycle testing) of isolated battery cells that have relatively low voltage and high capacity. There are also devices that allow testing of composite Li-ion batteries [9–10]. The disadvantages of the devices listed above include:

- input current of converter is not high enough: testing of batteries with rated capacity of 90 Ah requires current more than 100 A, which require using of multiple parallel connected devices;
- lack of opportunity to discharge battery to a negative voltage (polarity reversal), which does not allow to study the operation of the battery cells in the emergency situations..

2. Charge-discharge unit description

An automated CDU [11–12] (fig.1.) allows performing a full cycle of electrical tests.

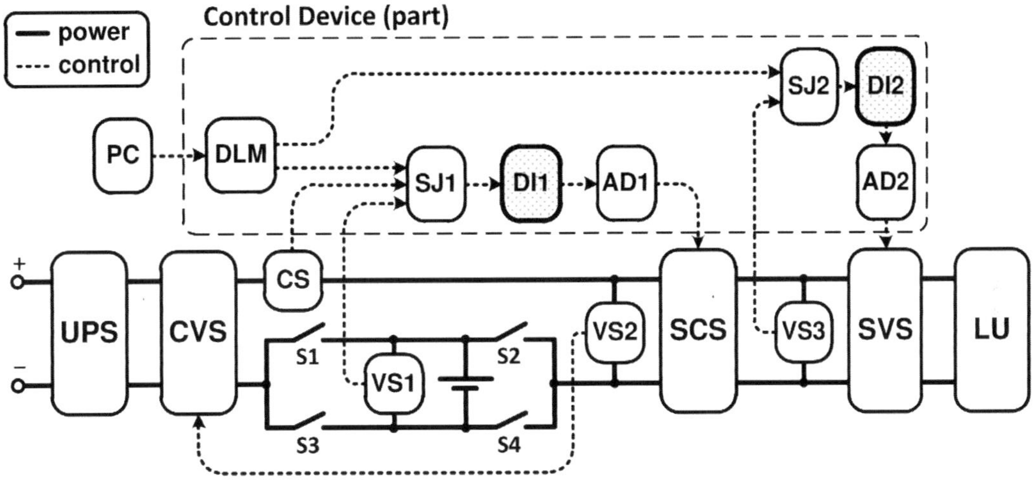

Figure 1. Block diagram of charge-discharge unit (CDU)

Uninterruptible power supply (UPS) prevents interruption of the tests in the case of emergency AC network shutdown. Charging and discharging modes of the battery are provided by appropriate position of switches S1-S4 and by a controlled voltage stabilizer (CVS). CVS allows ensuring stable charge/discharge mode operation. Excess of energy, consumed by CDU, is dissipated by a load unit (LU).

Necessity of current regulation in the wide range (0..160 A), the need for electrical isolation between switches S1-S4 and LU and complexity of high-frequency power transformer made it necessary to use two more stabilizers switched-mode stabilizer current (SCS) and switched-mode stabilizer voltage (SVS). SCS stabilizes its own input current, SVS stabilizes the SCS output voltage. SCS consist of switched-mode current regulator, which is based on boost converter. SVS consist of switched-mode voltage regulator, which is based on full-bridge buck converter.[13-14]

In order to provide high currents at low battery voltages it is necessary to maintain the SVS input voltage at the relatively high level. For this reason the stabilization of the SCS input voltage is provided by CVS (voltage sensor VS2 forms feedback for CVS). Therefore, regardless of battery voltage and the operating mode (charging or discharging of the battery) CVS stabilizes the SCS input voltage at the level of 4 V. In addition, the stabilization of the SCS input voltage enables to discharge the battery until polarity reversal.

A control device (CD), in accordance with predetermined patterns and the measured values of battery voltages and currents, controls switches S1-S4 (control routes between CD and switches are not shown on the diagram), generates control signals for SCS and SVS, and provides safe completion of tests if charge of the UPS batteries is low and if voltage in AC network is absent. CD includes digital logic module (DLM) which connected to personal computer (PC) and forms references for SCS and SVS summing junctions (SJ) on which they are added to the signals from the current (CS) and voltage (VS1, VS3) sensors. Control signals for SCS and SVS are filtered by digital integrators (DI) (detailed description will be given below) and amplified by analog drivers (AD).

CDU performs following functions:

- stabilization and regulation of charge and discharge currents;
- stabilization and regulation of discharge power;
- reproduction of electrical and temporal operational modes of battery;
- automatic diagnostics of CDU condition;
- emergency operation protection.

Document [15] gives guidelines for the definition of performance characteristics, guidelines for Li-ion battery cycle testing, and recommended measurement errors for battery parameters which are:

a) ± 0.1% for voltages;
b) ± 1% for currents;
c) ± 0.1% for temporal parameters.

At the same time, document [15] does not establish requirements for the stabilization accuracy of charge and discharge currents and discharge powers. In accordance with the adopted in the design of control systems practice, take attributes stabilization errors equal to measurement errors:

a) ± 1% for currents;
b) ± 0.1% for temporal parameters;
c) ± 1.1% for power.

Providing the required current measurement accuracy is achieved by using high-precision current sensor and by current sensor calibration. Calibration of current sensors performs at the intervals specified in the automatic battery testing program and automatically before startup. Calibration allows reducing to a minimum the statistical component of the current sensor error.

Measurement of the battery voltage and voltages of current sensors is performed using specialized measuring device based on National Instruments modules. Since the electrical isolation of measurement channels is required, NI PXIe-4300 modules are used [16]. They provide 8 channels, which allow simultaneously measuring, and have percentage errors of measurement no more than 0.0245%.

Stabilization and control of the attributes is based on digital signal processing in the single-chip computer. Accuracy of stabilization consists of the sum of the errors:

- the current and voltage measurement errors (discussed above);
- errors caused by the analog-digital converter (ADC);
- errors caused by of the reference voltage source;
- errors of attributes stabilization control loops.

Errors of the ADC and reference voltage are minimized by selecting appropriate electronic components. Charge-discharge unit, described by the authors in [11-12], contains a switching regulator of the input current (SRIC) and a switching regulator of the SRIC output voltage (SRSOV).

3. Mathematical model

The mathematical model of the power part of CDU is based on the block diagram (fig. 1) and the equivalent circuit (fig. 2). This model allows analyzing the accuracy of CDU static stabilization.

Variations of voltages and currents in circuits are negligible in steady state, which allows omitting the parameters of reactive elements in the mathematical model (inductances and capacitances). A linearized system of steady-state equations (1) describes relations between currents and voltages in the circuit shown in fig. 2:

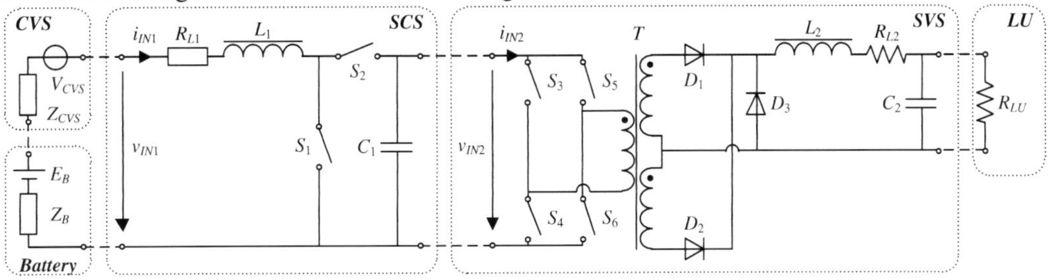

Figure 2. Equivalent electrical circuit of the CDU power part

$$\begin{cases} V_S = V_{CVS} + E_B \\ v_{IN1} = V_S - i_{IN1}(R_{CVS} + R_B) \\ i_{IN1} = I_{IN1}D_I + I_{IN1}d_i + i_{in1}D_I + i_{IN2} \\ v_{IN2} = v_{IN1} - i_{IN1}R_{L1} + V_{IN2}D_I + v_{in2}D_I + V_{IN2}d_i \, , \\ i_{IN2} = (I_{LU}D_V + i_{lu}D_V + I_{LU}d_v)n \\ v_{LU} = (V_{IN2}D_V + v_{in2}D_V + V_{IN2}d_v)n - i_{LU}R_{L2} \\ i_{LU} = v_{LU}G_{LU} \end{cases} \tag{1}$$

where V_S – voltage of the source, V_{CVS} – dc voltage at CVS output, E_B – battery cell dc voltage, v_{IN1} –SCS input voltage, $i_{IN1} = I_{IN1} + i_{in1}$ – total value of SCS input current, I_{IN1} – equilibrium point value of i_{IN1}, i_{in1} – variation of i_{IN1}, R_{CVS} – CVS resistance, R_B – battery cell resistance, $d_I = D_I + d_i$ – total value of SCS duty cycle, D_I – equilibrium point value of d_I, d_i – variation of d_I, i_{IN2} – SVS input current, $v_{IN2} = V_{IN2} + v_{in2}$ – total value of SVS input voltage, V_{IN2} – equilibrium point value of v_{IN2}, v_{in2} – variation of v_{IN2}, R_{L1} – resistance of L_1, i_{IN2} – SVS input current, $i_{LU} = I_{LU} + i_{lu}$ – total value of LU current, I_{LU} – equilibrium point value of i_{LU}, i_{lu} – variation of i_{LU}, $d_V = D_V + d_v$ – total value of SVS duty cycle, D_V – equilibrium point value of d_V, d_v – variation of d_V, n – turns ratio of transformer T, v_{LU} – LU voltage, R_{L2} – resistance of L_2, G_{LU} – conductance of LU.

The system of equations (2) describes a control system of CDU. Along with (1), it forms a mathematical model of CDU. Block diagram of total model (fig. 3) consists of systems (1) and (2):

$$\begin{cases} d_I = (V_{REFI} - i_{IN1}K_{CS})K_I K_{PWMI} \\ d_V = (v_{IN2}K_{VS} - V_{REFV})K_V K_{PWMV} \end{cases} , \tag{2}$$

where V_{REFI} – reference voltage of SCS feedback loop, K_{CS} – current sensor gain, K_I – gain of SCS feedback loop, K_{PWMI} – gain of SCS pulse width modulator (PWM), V_{REFV} – reference

IOP Conf. Series: Materials Science and Engineering **255** (2017) 012016 doi:10.1088/1757-899X/255/1/012016

voltage of SVS feedback, K_{VS} – voltage sensor gain, K_V – gain of SVS feedback loop, K_{PWMV} – gain of SVS PWM.

Expressions for the current and voltage with offset that appeared as a result of the presence of steady-state error were obtained from systems (1) and (2):

$$i_{IN1} = f(V_S, V_{REFI}, V_{REFV}, K_I, K_V), \tag{3}$$

$$v_{IN2} = f(V_S, V_{REFI}, V_{REFV}, K_I, K_V). \tag{4}$$

Open-loop transfer functions of SCS and SVS control loops at steady state become SCS gain KOLI and SVS gain KOLV, respectively, which are functions of the following variables:

$$K_{OLI} = f(V_S, V_{REFI}, V_{REFV}, K_I, K_V), \tag{5}$$

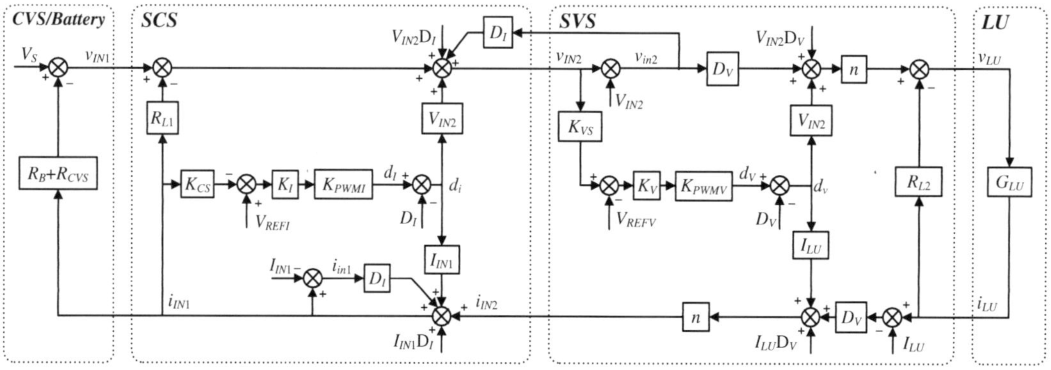

Figure 3. Block diagram of linearized CDU model for steady-state conditions

$$K_{OLV} = f(V_S, V_{REFI}, V_{REFV}, K_I, K_V). \tag{6}$$

Since the CDU input current varies in a wide range of values down to zero, usage of the standard relative error for evaluation of the system accuracy will give to large errors at low currents. Percentage error normalized relative to maximum measurement value (hereinafter – percentage error, designated as δ_M) is more convenient for evaluation purposes [17–18]. The percent error of SCS control loop δ_{MI} is used for SCS input current stabilization accuracy estimation:

$$\delta_{MI} = \left| \frac{I_{IN1} - i_{IN1}}{I_{IN1\max}} \right|, \tag{7}$$

where $i_{IN1} = I_{IN1} + i_{in1}$ – total value of SCS input current, I_{IN1} – equilibrium point value of i_{IN1}, I_{IN1max} – the maximum current of the CDU (I_{IN1max} = 160 A).

SCS percentage error δ_{MI} for the CDU was set to 0.5%.

The percent error of SVS control loop δ_{MV} is similarly used for SVS input voltage stabilization accuracy estimation:

$$\delta_{MV} = \left| \frac{V_{IN2} - v_{IN2}}{V_{IN2\max}} \right|, \tag{8}$$

IOP Conf. Series: Materials Science and Engineering **255** (2017) 012016 doi:10.1088/1757-899X/255/1/012016

where $v_{IN2} = V_{IN2} + v_{in2}$ – total value of SVS input voltage, V_{IN2} – equilibrium point value of v_{IN2}, V_{IN2max} – the maximum input voltage of SVS.

4. Simulation and experimental results

Study the impact of various operational modes of the CDU on the SCS current percentage error allows determining the device parameters that ensure the required accuracy of the CDU input current stabilization. Modes of operation of the CDU are defined by different values of the following quantities: the input current I_{IN1}, voltage of the source V_S, SCS gain K_{OLI} and SVS gain K_{OLV}.

Equations (3)-(8) allow obtaining the implicit dependence of SCS δ_{MI} and SVS δ_{MV} percentage errors from SCS K_{OLI} and SVS K_{OLV} gains (fig.4-9).

Figures 4 and 5 show that with an increase in SCS gain K_{OLI} value of SCS percentage error δ_{MI} decreases, regardless of the action of various destabilizing factors.

Figure 6 shows that SCS percentage error δ_{MI} remains practically constant with increasing of SVS gain K_{OLV} and decreases with increasing of SCS gain K_{OLI}.

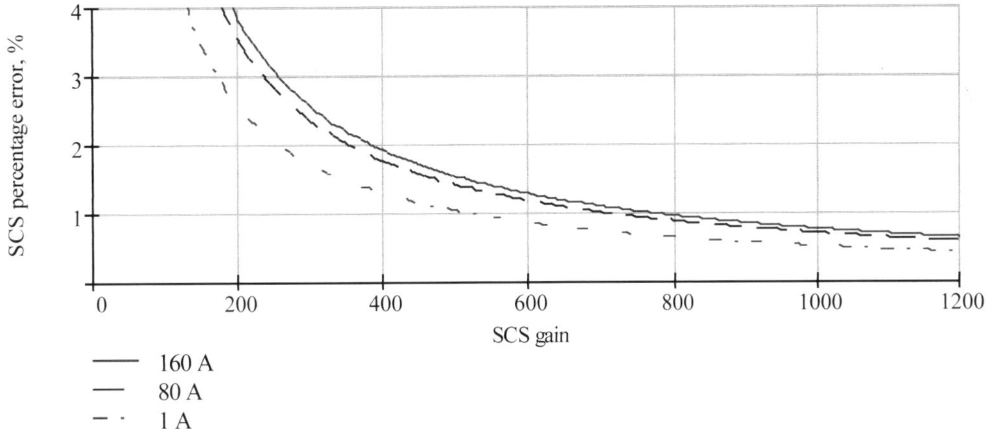

Figure 4. An implicit relation between SCS percentage error δ_{MI} error and SCS gain K_{OLI} for different values of the input current I_{IN1}

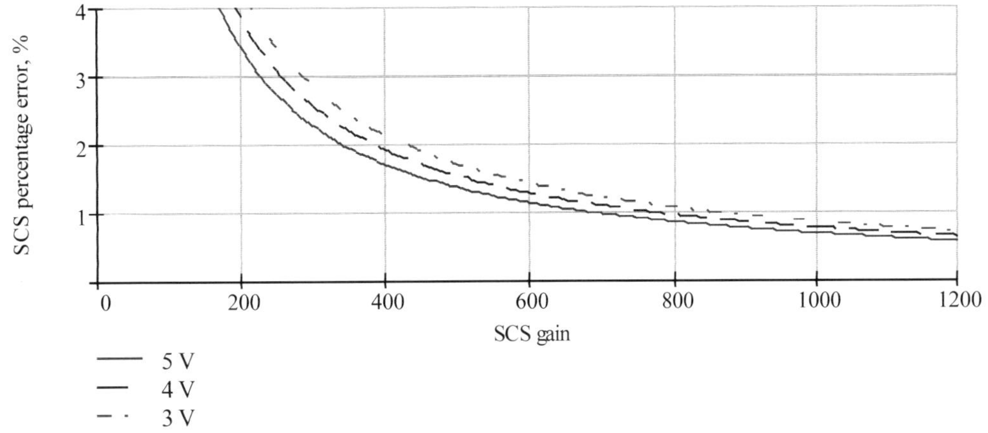

Figure 5. An implicit relation between SCS percentage error δ_{MI} and SCS gain K_{OLI} for different values of the input voltage V_S

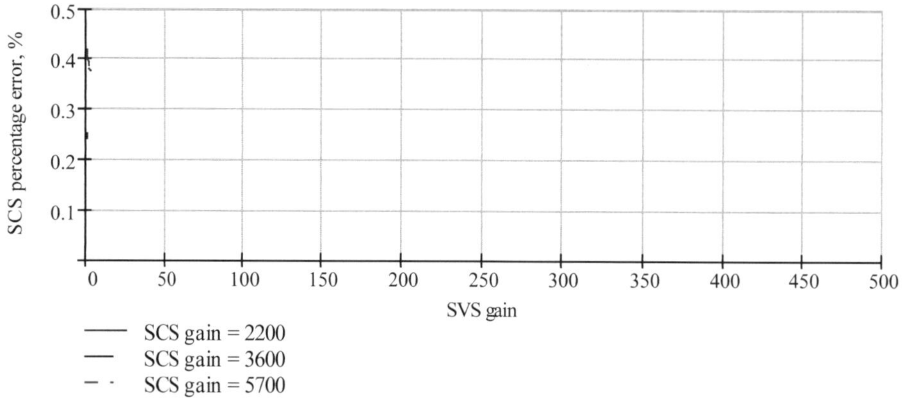

Figure 6. An implicit relation between SCS percentage error δ_{MI} and SVS gain K_{OLV} for different values of SCS gain K_{OLI}

The value of SVS gain K_{OLV} determines the SVS percentage error δ_{MV}. It was established experimentally that for the normal CDU operation a value of SVS percentage error δ_{MV} should be 3-5% over the entire range of input voltages and currents. Figures 7-9 show that to ensure this requirement is necessary that the value of SVS gain K_{OLV} was big enough.

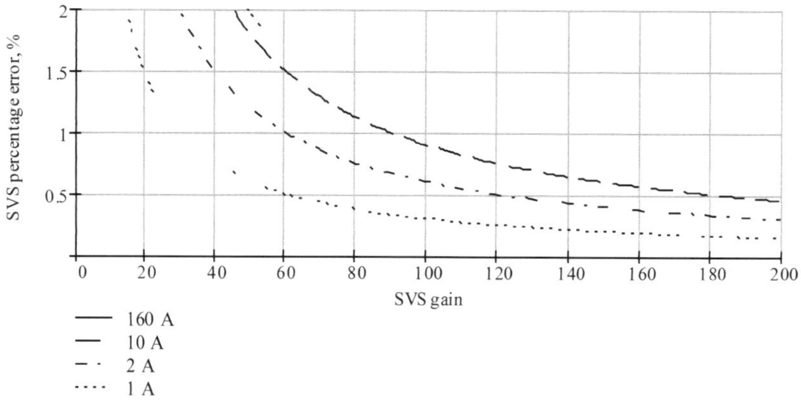

Figure 7. An implicit relation between SVS percentage error δ_{MV} and SVS gain K_{OLV} for different values of the input current I_{IN1}

Figures 7 and 8 show that SVS percentage error δ_{MV} decreases with increasing SVS gain K_{OLV} and does not depend on the input voltage V_S.

Figure 9 shows that SVS percentage error δ_{MV} significantly depends on SCS gain K_{OLI} and the input current I_{IN1}, which imposes restrictions on the minimum value of SCS gain K_{OLI}.

It can be concluded from the results of mathematical modeling that SCS gain K_{OLI} and SVS gain K_{OLV} are interrelated. It is methodically reasonable to select the value of SVS gain K_{OLV} first and then to select large enough SCS gain K_{OLI}.

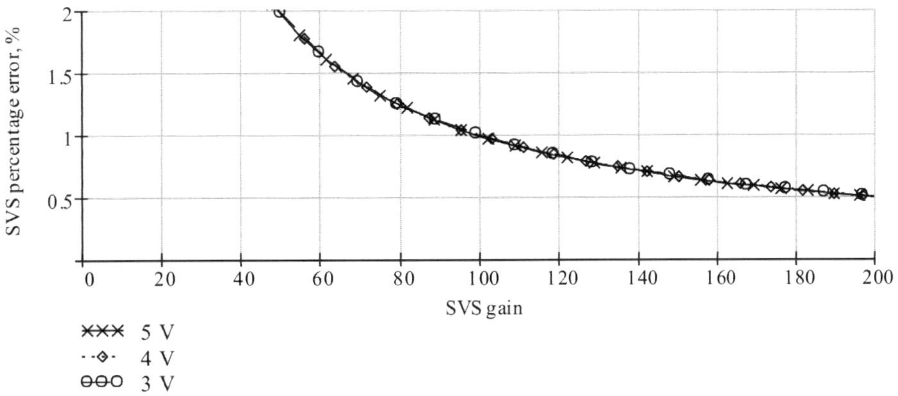

Figure 8. An implicit relation between SVS percentage error δ_{MV} and SVS gain K_{OLV} for different values of the input voltage V_S

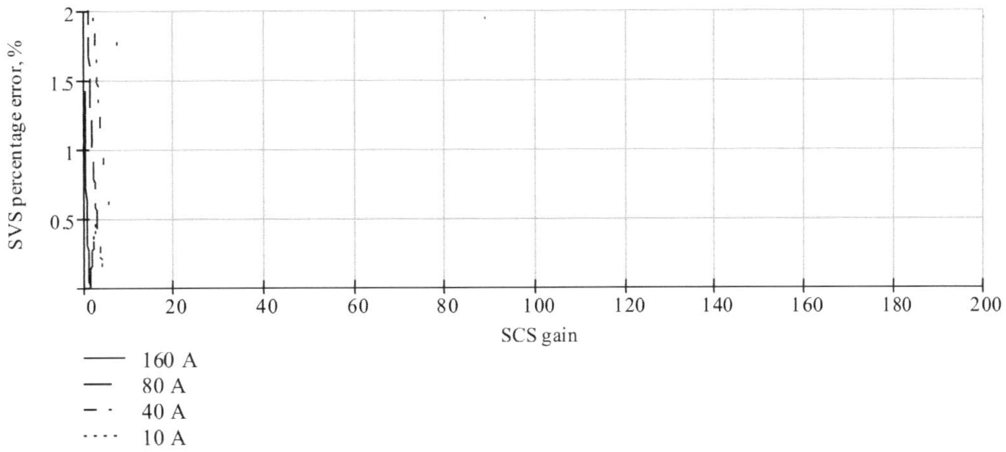

Figure 9. An implicit relation between SVS percentage error δ_{MV} and SCS gain K_{OLI} for different values of the input current I_{IN1}

The required accuracy of attribute stabilization was also achieved by increasing system type by least 1 through the introduction of the digital integrator in the SCS control loop [19].

A difference equation of the digital integrator is of the form:

$$E_g[n] = k \cdot E_g[n] + E_g[n-1], \tag{9}$$

where n – unitless time, k – integration ratio.

To illustrate the correctness of the developed approach for CDU parameters, experimental studies of SCS percentage error for LISP-85 Li-ion battery [20] were performed (fig. 10).

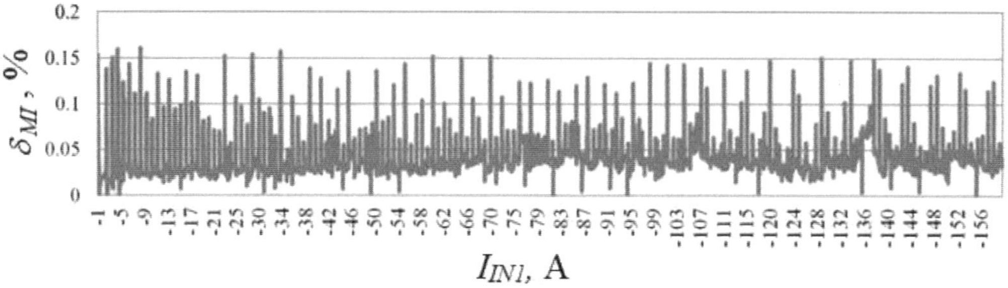

Figure 10. Experimental relation between SCS percentage error δ_{MI} and depending on input current I_{IN1}

The experimental results confirm that the SCS percentage error does not exceed 0.2%, which is consistent with the selected design conditions ($\delta_{MI} = 0.5\%$).

5. Conclusion

The proposed approach to the selection of the CDU parameters provides the desired percentage errors of required attributes of Li-ion battery operational modes. SCS gain K_{OLI} and SVS gain K_{OLV} are interrelated, which can be confirmed by the results of mathematical modelling. For this reason, it makes sense to select the value of SVS gain K_{OLV} first and then to select large enough SCS gain K_{OLI}. The required accuracy of attribute stabilization was also achieved by increasing system type by least 1 through the introduction of the digital integrator in the SCS control loop.

Acknowledgements

This study was supported by the Ministry of Education and Science of the Russian Federation (Government Contract 14.577.21.0082, unique identifier RFMEFI57714X0082).

References

[1] Yang F Y, Xing Y, Wang D, Tsui K-L 2016 A comparative study of three model-based algorithms for estimating state-of-charge of lithium-ion batteries under a new combined dynamic loading profile *Applied Energy* vol **164** pp 387–399 doi: 10.1016/j.apenergy.2015.11.072

[2] Hussein A A 2015 Capacity Fade Estimation in Electric Vehicle Li-Ion Batteries Using Artificial Neural Networks *IEEE Trans. on Industry Applications* vol **51** No. 3 pp 2321–2330 doi: 10.1109/TIA.2014.2365152

[3] Xing Y, He W, Pecht M, Tsui K L 2014State of charge estimation of lithium-ion batteries using the open-circuit voltage at various ambient temperatures *Applied Energy* vol **113** pp 106–115 doi: 10.1016/j.apenergy.2013.07.008

[4] Sight H N 2009 Lithium-ion battery prognostic testing and process. Patent US, No. 7576545 B2,

[5] Kwok W Y 2001 Method of calculating dynamic state-of-charge within a battery. Patent US, No. 6300763 B1

[6] Guiheen J V, Sight H, Palanisamy T G 2003 Method and apparatus for determining the state of charge of a lithium-ion battery. Patent US, No. 6586130 B1

[7] Sayfang G R 1990 Battery state of charge indicator. Patent US, No. 4949046 A

[8] Palanisamy T, Sight H, Atehortua H, Hoenig S 2005 Method and apparatus for determining the available of energy of a lithium ion battery. Patent US, No. 20050269993 A1
[9] Chroma Systems Solutions, Inc. 17011 Programmable Charge/Discharge Tester. Available at: http://www.chromausa.com/product/17011-programmable-chargedischarge-tester/ (accessed 01.02.2017)
[10] Cadex Electronics Inc. C8000 BATTERY TESTING SYSTEM. Available at: http://www.cadex.com/en/products/c8000-battery-testing-system/ (accessed 01.02.2017)
[11] Kopylov E A, Mizrah E A, Fedchenko A S, Lobanov D K 2016Study of a lithium-ion battery charge-discharge test unit characteristics *IOP Conf. Series: Materials Science and Engineering* vol **122** doi: 10.1088/1757-899X/122/1/012015
[12] Fedchenko A S, Kopylov E A, Lobanov D K, Mizrah E A 2016 Static accuracy of the automated stand for lithiumion batteries testing *18th Mediterranean Electrotechnical Conference (MELECON)*, Lemesos doi: 10.1109/MELCON.2016.7495336
[13] Gu B, Lai J-S, Kees N, Zheng C 2013 Hybrid-Switching Full-Bridge DC–DC Converter With Minimal Voltage Stress of Bridge Rectifier, Reduced Circulating Losses, and Filter Requirement for Electric Vehicle Battery Chargers *IEEE Trans. on Power Electronics* vol **28** No. 3 pp 1132–1144 doi: 10.1109/TPEL.2012.2210565
[14] Gautam D ., Musavi F, Eberle W, Dunford W G 2013 A Zero-Voltage Switching Full-Bridge DC--DC Converter With Capacitive Output Filter for Plug-In Hybrid Electric Vehicle Battery Charging *IEEE Trans. on Power Electronics* vol **28** No. 12 pp 5728–5735 doi: 10.1109/TPEL.2013.2249671
[15] GOST R MEK 62660-1-2014. Akkumulyatory litiy-ionnye dlya elektricheskikh dorozhnykh transportnykh sredstv. Chast' 1. Opredelenie rabochikh kharakteristik. [IEC 62660-1:2010. Secondary lithium-ion cells for the propulsion of electric road vehicles. Part 1. Performance testing]. Moscow, Standartinform Publ., 2015. 34 p.
[16] National Instruments Corporation PXIe-4300. Available at: http://www.ni.com/ru-ru/support/model.pxie-4300.html (accessed 01.02.2017)
[17] Mizrah E A 2001 Method for assessing the accuracy of spacecraft primary energy source simulators *Vestnik SibGAU* vol **8** No. 2 pp 10–14 (In Russ.)
[18] Mizrah E A 2005 The analysis of static accuracy of solar batteries simulators *Vestnik SibGAU* vol **12** No. 4 pp 24–27 (In Russ.)
[19] Selvaraj J, Rahim N A 2008 Multilevel Inverter For Grid-Connected PV System Employing Digital PI Controller *IEEE Trans. on Industrial Electronics* vol **56** No. 1 pp 149–158 doi: 10.1109/TIE.2008.928116
[20] JSC «Saturn» LISP-85. Available at: http://www.saturn.kuban.ru/liab_spec.html (accessed 01.02.2017)

IOP Conf. Series: Materials Science and Engineering **255** (2017) 012017 doi:10.1088/1757-899X/255/1/012017

Optimizing the control process parameters for the induction soldering of aluminium alloy waveguide paths[1]

V S Tynchenko, A V Murygin, V E Petrenko, O A Emilova, A N Bocharov
Reshetnev Siberian State University of Science and Technology, 31, Krasnoyarsky Rabochy Av., Krasnoyarsk, 660037, Russian Federation

E-mail: vadimond@mail.ru

Abstract. The paper describes the problem of selecting the optimal initial values of algorithm parameters for the process of the induction soldering of aluminium alloy waveguide paths. The authors consider some factors influencing the quality of waveguide soldered joint elements. These factors depend on the correct choice of initial values for control parameters. The problem of optimizing such parameters for further analytical and numerical studies is researched by authors. For solving the stated task, the random search method is selected, allowing for an acceptable field study within the stated time to solve the problem of control optimization with the level of accuracy required by the technological process. Therefore, optimal initial values of the induction soldering technological process were found for three sizes of waveguide tubes and flanges.

1. Introduction

The induction soldering of aluminium alloy waveguide paths is a high-tech process that requires complex modern automated equipment and the highly accurate refinement of parameter values [1, 2, 3]. The main difficulties consist in the diversity of the possible sizes of products, and also in the need to fine-tune the technological process for each of them, as well as in the presence of an inductor with a complex profile [4, 5, 6, 7].

The authors of this work implemented a set of automated equipment, allowing the process of soldering waveguide paths to be performed. This process is controlled by the preinstallation of a gap by means of an "inductor-flange" with a margin of error of no more than 0.2 mm, and for each product empirical values are required matching the clearance. Then the technological process of soldering is launched in the control system, during which the generator power is regulated in accordance with the established heating rate and stabilization mode settings when the melting temperature of the solder is reached. [8, 9, 10]

However, if such a control system is used, there is the problem of refining the clearance, whereby even small deviations can lead to breaches of the conditions for obtaining high-quality connections. As a result, the process management system operator is forced to adjust

[1] The reported study was funded by Russian Foundation for Basic Research, Government of Krasnoyarsk Territory, Krasnoyarsk Region Science and Technology Support Fund to the research project № 16-48-242029

the clearance of the inductor-flange, thereby redistributing the currents induced in the workpiece in the following situations [11]:

- if the process reaches the melting temperature of the solder leading to the convergence of the process;
- if the process encounters a situation whereby it is obvious that there is a significant deviation from the normal operation of the system;
- failover process.

In consequence, the human factor influencing the quality of finished product is crucial, and in this way, the percentage of defects depends on the experience of the induction soldering process operator. This results in the need to modernize the automated equipment in order to exclude the influence of the human factor, to accelerate the technological process and to improve the repeatability of the process.

2. Description of the specifics of the waveguide path soldering process

After modernizing the equipment and performing preliminary tests, there arose the problem of optimizing the process with the new increased complexity of the system. The insertion of a second control circuit that changes the position of the workpiece relative to the inductor window during the soldering process made the system multiply connected, nonlinear and nonstationary. This issue was resolved through the development of a logical control regulator, which allowed the soldering process to be controlled regardless of the size of the soldered element with a wide range of technological process initial parameters.

Initial launch parameters have become crucial in ensuring the required regulation performance of the system, as the regulator settings are selected in such a way as to guarantee proper control of the soldering process with elements of any size. Such an influence corresponds to a high level of dependence on the initial temperature distribution tubes and flange/coupling, due to the high threshold of the perception pyrometric temperature sensor, which is 300 degrees Celsius.

However, a non-optimized soldering process allows us to get soldered connections by reducing the starting power, which significantly increases the time the process takes, diminishing the electromagnetic characteristics of the finished product. Also, a strong variation of the initial temperature distribution between the elements of the soldered product can have such values, under which subsequent regulation fails to reduce the heating process to the required technological parameters for the subsequent limited plot.

3. The problem of choosing the control parameters

The complexity of the algorithm in the automated system with a second control circuit leads to the appearance of additional control unit settings. The heating process quality is significantly influenced by initial control system settings, such as:

- the initial power fed to the inductor;
- the initial position of the workpiece relative to the inductor window.

When using a single-loop control, the starting power was set to the maximum possible for the generator, because reducing the temperature difference between the waveguide elements was not the goal of the regulation, and the distance between the waveguide and inductor window was constant throughout the technological process of soldering. Now with double-circuit control the aim is to improve the quality of the technological process by eliminating temperature differences between elements of soldered product. Due to this there is a problem of choosing the initial value of the power supplied to the inductor, as well as the clearance between the workpiece and the inductor window.

In a separate work, the authors proposed and tested methods and algorithms to control the waveguide path induction soldering process using a two-circuit control system for the temperature of the elements of the product to be soldered and the gap between the inductor window and the connecting flange/coupling.

The present study task is to select the initial values of the studied parameters of the control system.

To solve such a problem, it is necessary to determine the correlation between the adoption of parameters of boundary values and the resulting influence on the heating process and the quality of the soldered connection.

Table 1 summarizes the results of experimental research on the initial parameter values.

An example that illustrates the implications of choosing a high initial value for the power generator is presented in Figure 1.

In the process presented in Figure 1, a high weakly damped oscillation occurs in the control process. Also, there are examples of swelling and melting of the main material.

Table 1. Correspondence of control system parameter initial values and their influence on the quality of the induction soldering technological process

Parameter	Meaning	Consequences
Generator power	High	Local overheating of the product, swelling and melting of the main material (for couplings). A high initial spread of the temperature distribution of the product elements. The need to search for the initial position and high-precision positioning of the workpiece relative to the inductor window to obtain an even distribution of heating between the elements of the product.
	Low	Time delays in the process. Reductions in the efficiency of the generator due to heat loss during convection and self-heating of the inductor. Ambiguity in the choice of the initial position of the workpiece to the inductor window. When registering temperatures, the flange is superheated, which subsequently leads to its melting.
The gap between the inductor and the product	High	Overheating of the tube outside the pyrometer temperature control zone. An essentially uneven temperature field across the product section. Poor warming of the flange, leading to a tightening of the soldering process. Long-term transient when regulating. The probability of reflow of the tube's top and local burns of the main material at the end of the soldering process.
	Low	Overheating of the flange in the preheating stage, resulting in a long transient during adjustment. High probability of flashing at the end of the soldering process.

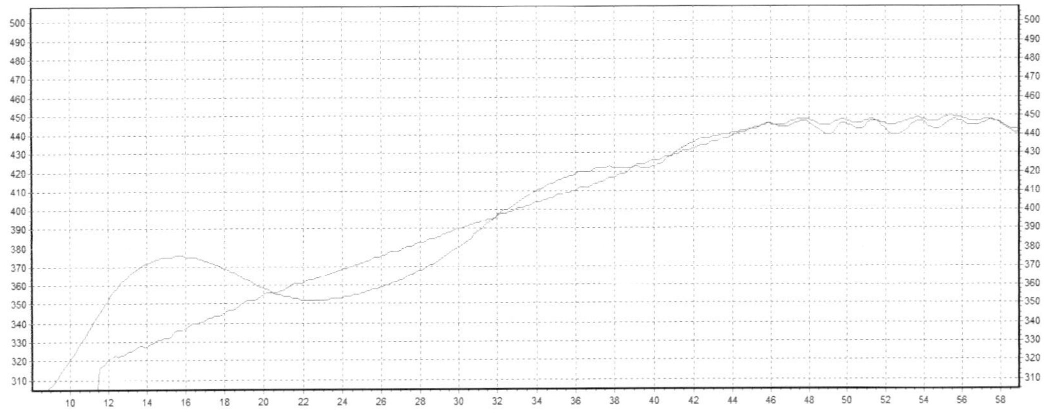

Figure 1. Temperature charts of elements with an initial high power value of the power generator
Red line – tube temperature, blue line - temperature of the flange.
Y axis - temperature, X axis - time.

Examples to illustrate the implications of choosing extreme values for the gap between the inductor and the product are presented in Figures 2 and 3.

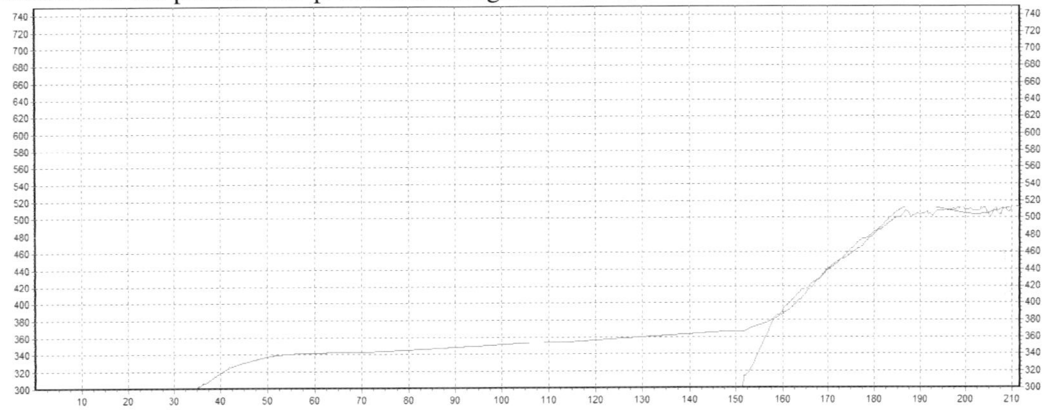

Figure 2. Graph showing the heating tube-flange (58 × 25 mm) with the initial gap between the inductor and flange of 3 mm
Red line – tube temperature, blue line - temperature of the flange.
Y axis - temperature, X axis - time.

In the process presented in Fig. 2 can be seen significant overheating of the flange relative to the tube at the beginning of the technological process. The total time for the soldering process is protracted. In the area of solder melting (520-540 Celsius degrees), there is a weakly damped oscillation. On the product elements occur swelling and melting of the core material.

In the process presented in Fig. 3 can be seen significant overheating of the tube relative to the flange at the beginning of the technological process. The total time for the soldering process is protracted. On the process stabilizing phase (solder melting point) the heat distribution in the tube leads to it being impossible for the regulator to reduce the difference in product element temperatures.

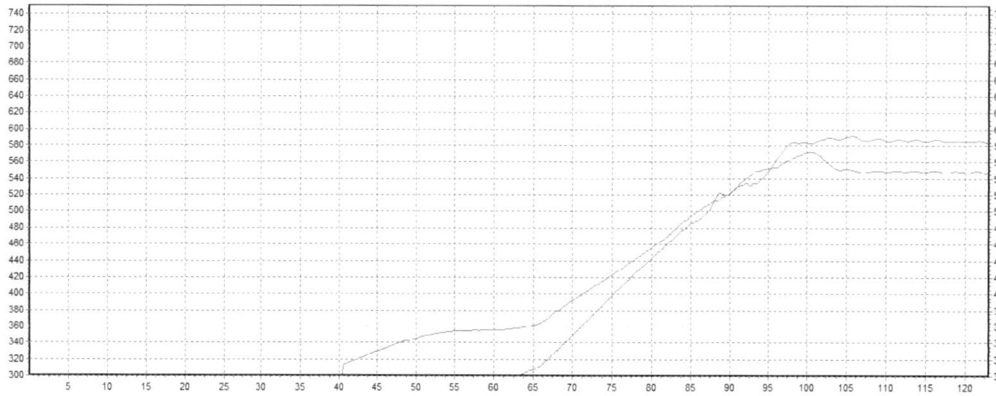

Figure 3. Graph showing the heating tube-flange (58 × 25 mm) with the initial gap between the inductor and flange of 9 mm
Red line – tube temperature, blue line - temperature of the flange.
Y axis - temperature, X axis - time.

4. The estimation of the induction soldering control quality

The need to select optimal values for the control system initial parameters can be explained by such reasons as:

- flux activity time after its meltdown (490-520 degrees Celsius) is about 10 seconds, which leads to the necessity of the heating speed being at a minimum of 8 degrees Celsius per second from the flux melting until the solder melting;
- selecting the workpiece from an incorrect starting position relative to the inductor window, the control system fails to reduce the difference between temperatures of the product elements according to the required technological standards;
- proper selection of the control process initial parameters allows quality soldered connections of waveguide elements to be formed within a reasonable time.

Quality control can be estimated as the integral of the tube and flange/coupling temperature difference function (1):

$$f(W,h) = \int_0^{\infty} \left(T_{fl}(t,W,h) - T_{tb}(t,W,h)\right) dt \rightarrow \min_{\substack{W>0 \\ h>0}} \quad (1)$$

where W is the generator power;

h is the gap size;

$T_{fl}(t,W,h)$ is the flange/coupling temperature;

$T_{tb}(t,W,h)$ is the tube temperature.

To use numerical optimization methods instead of the target function (1), the average quadratic deviation between the temperatures of the heated elements can be employed.

Then the target feature that provides the reduction of the difference in temperature of soldered elements takes the form (2):

$$F(W,h) = \frac{1}{n}\sqrt{\sum_{i=1}^{n}\left(T_{fl}^{i} - T_{tb}^{i}\right)^{2}} \rightarrow \min_{\substack{W>0\\h>0}} \quad (2)$$

where W is the generator power;

h is the size of the gap;

n is the number of function evaluations;

T_{fl}^{i} is the measured temperature of the flange/coupling at the i-th point of the technological process;

T_{tb}^{i} is the measured temperature of the tube at the i-th point of the technological process.

5. Methods of problem solving

There are a variety of methods to solve the optimization task:

1) direct methods;
2) gradient methods;
3) Newton's methods.

Because neither the task of optimizing the soldering technological process parameters, nor the research object has a well-defined mathematical model, and the value of the function is obtained by conducting full-scale welding experiments on the product test samples, only methods belonging to the first group, namely direct methods, can be used to solve the optimization problem of such a process, those that use only information about the target function values.

The problem of this study has several characteristic features:

1) low dimensionality of the search space;
2) small uncertainty intervals;
3) optimized parameters with low discreteness;
4) a limited number of possible target function calculations due to the high cost of the full-scale experiment.

Among the group of direct methods, the most commonly used are [12]:

1) the uniform search method;
2) the Golden section method;
3) the Fibonacci method;
4) the random search method in its different variations;
5) Powell's method;
6) the method of Hooke-Jeeves;
7) the method of Nelder-Mead.

The first three methods find wide application in problems of one-parameter optimization, however, are poorly suited to multidimensional cases.

The methods of Powell, Hooke-Jeeves and Nelder-Mead require quite a large amount of target function computing to implement at least one iteration of the algorithm, which limits their application in the field of experimental research. The random search method is widely used in practical research with an unknown object or process being analysed [13]. Thus, the random search method was selected to be used for the parameter optimization of the process control system for waveguide path induction soldering.

The random search method has several different variations:

1) the size and step at each iteration method can be a constant or chosen individual for each iteration;

2) the possibility to return after a bad step can be present or absent;

3) the choice of coordinates whose value should be changed on iteration can be made randomly or on the basis of additional obtained information.

The following random search method modification was selected in this study:

1) a different step value is selected by the researcher for each iteration, due to the active direct participation of researchers in the process of parameter optimization;

2) return after a bad step;

3) the search direction is chosen at random due to limited information about the process.

6. Experimental study

Using the random search method, a selection of the optimal values for the technological process initial parameters was made for the following waveguide sizes:

1) 58 × 25 mm;

2) 35 × 15 mm;

3) 19 × 9.5 mm.

The search for optimal parameters ends when the values of the target functions reached 5 degrees Celsius, due to technological regulations.

Table 2 provides step-by-step solution results for the optimization problem of searching for optimal parameters for a waveguide with a size of 58 × 25 mm. The solution is marked in bold. Figure 4 illustrates the search process in the coordinate space of the optimized parameters.

Table 2. Step-by-step solution results for the optimization problem for size 58 × 25 mm

Iteration of the algorithm	Power (W), kW	The size of the gap (h), mm	The action of the algorithm	The value of the objective function, degrees Celsius
1	5	5	Beginning	30.7
2	5	6	Transition	24.3
3	7	6	Transition	21.1
4	9	6	Transition	16.4
5	9	5.5	Transition	10.7
6	**10**	**5.5**	**End of the search**	**2.5**

Table 3 provides step-by-step solution results for the optimization problem of searching for optimal parameters for a waveguide with a size of 35 × 15 mm. The solution is marked in bold. Figure 5 illustrates the search process in the coordinate space of the optimized parameters.

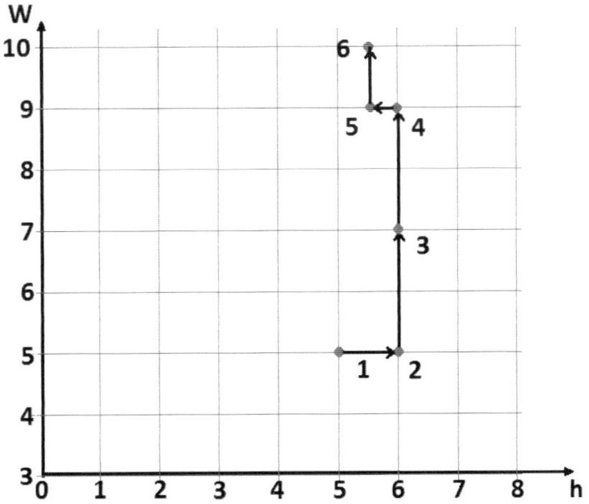

Figure 4. The search process for the optimal parameters for a standard size 58 × 25 mm

Table 3. Step-by-step solution results of optimization problem for size 35 × 15 mm

Iteration of the algorithm	Power (W), kW	The size of the gap (h), mm	The action of the algorithm	The value of the objective function, degrees Celsius
1	5	3	Beginning	21.5
2	6	3	Transition	17.6
3	7	3	Transition	9.7
4	8	3	Return	15.1
5	**7**	**4**	**End of the search**	**4.1**

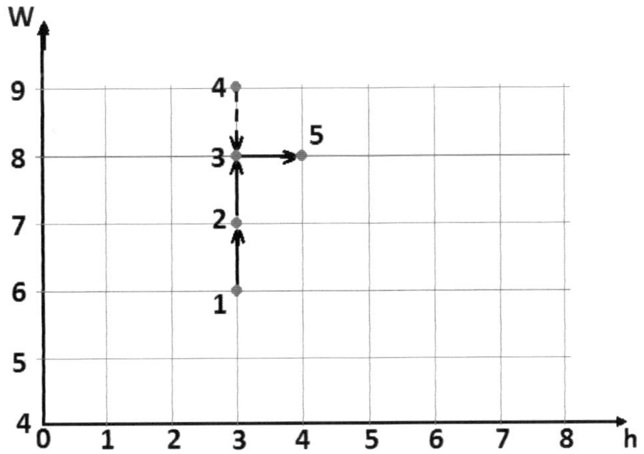

Figure 5. The search process for the optimal parameters for a standard size 35 × 15 mm

Table 4 provides step-by-step solution results for the optimization problem of searching for optimal parameters for a waveguide with a size of 19 × 9.5 mm. The solution is marked in bold. Figure 6 illustrates the search process in the coordinate space of the optimized parameters.

Table 4. Step-by-step solution results of optimization problem for size 19 × 9.5 mm

Iteration of the algorithm	Power (W), kW	The size of the gap (h), mm	The action of the algorithm	The value of the objective function, degrees Celsius
1	3	5	Beginning	42.4
2	3	4	Transition	23.7
3	3	3	Transition	17.1
4	3	2	Return	27.8
5	4	3	Transition	14.3
6	5	3	Transition	10.3
7	6	3	Return	18.7
8	**5**	**2.5**	**End of the search**	**3.8**

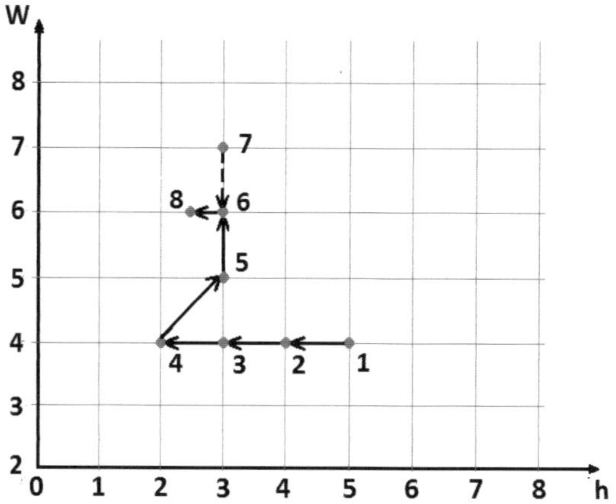

Figure 6. The search process for the optimal parameters for a standard size 19 × 9.5 mm

In Table 5 are shown the best initial values of the soldering process control parameter for different types and sizes of waveguides.

Table 5. - Optimal initial values of the soldering process control parameter

Size, mm	Power (W), kW	The size of the gap (h), mm
58x25	10	5.5
35x15	7	4
19x9.5	5	2.5

7. Discussion of results

From the data in Tables 2, 3 and 4 it is obvious that the application of the random search method allows the initial parameters for the control process of waveguide path induction soldering to be optimized under the conditions of a limited series of full-scale experiments.

The optimization of the explored waveguide size parameters did not demand more than eight iterations of the algorithm, which confirms the effectiveness of the target function selection (2) and random search method implementation to solve the stated problem.

The method allowed us to optimize settings for a wide range of types and sizes of waveguide paths, thus confirming the possibility of its application to the stated problem.

8. Conclusion

The following results were achieved in the study:

1) a form of target function for optimizing the initial parameters for the technological process of the induction soldering of aluminium alloy waveguide paths was proposed, and an effective method for parameter optimization was selected;

2) the best technological process initial parameters for the induction soldering were received for the standard sizes of 58x25 mm, 35x15 mm and 19x9.5 mm;

3) the effectiveness of using the random search method was confirmed, a method which can be used in the future to optimize control parameters for sizes that have not been investigated in the study.

The proposed decision allows the technological parameters to be optimized for the induction brazing process, reducing time, material and labour costs associated with the need to conduct full-scale experiments.

Thus, the practical significance of the research is confirmed.

References

[1] Zlobin S K, Mikhnev M M, Laptenok V D and Zaitsev R V 2012 Features of soldering waveguide distribution paths of aluminum alloys using induction heating source *Reshetnev reading : proceedings of the XVI Intern. scientific. conf.* vol **2** pp 192-193

[2] Laptenok V D, Zlobin S K, Mihnev M M and Bocharov A N 2013 Features of the production of the waveguide-junction of paths of antenna-feeder devices of spacecraft *Vestnik SibGAU* **6(52)** pp 196-201

[3] Laptenok V D, Dolgopolov B B and Bocharov A N 2009 The technique of debugging of technological process of soldering the waveguide paths (Krasnoyarsk : SibGAU) 112 p.

[4] Silchenko P N, Ankudinov A V and Kudryavtsev I V 2012 Ensuring the strength and accuracy of large-sized waveguide-distributive systems of space communication devices *Problemy mashinostroeniya i nadezhnosti mashyn* **1** pp 112-117

[5] Vologdin V V, Kushch E V, Assam V V 1989 Induction soldering (L. : «Mechanical engineering») 72 p

[6] Slugocki A E and Ryskin S E 1974 *Inductors for induction heating* (Leningrad : Energy)

[7] Slugocki A. E. 1989 Inductors (L.: Engineering)

[8] Murygin A V, Tynchenko V S, Laptenok V D, Emilova O A and Bocharov A N 2017 Complex of automated equipment and technologies for waveguides soldering using induction heating *IOP Conference Series: Materials Science and Engineering* **173**

[9] Tynchenko V S, Murygin A V, Emilova O A, Bocharov A N and Laptenok V D 2016 The automated system for technological process of spacecraft's waveguide paths soldering *IOP Conference Series: Materials Science and Engineering* **155**.

[10] Tynchenko V S, Bocharov A N, Laptenok V D, Seregin Yu N and Zlobin S K 2016 The software of technological process of spacecraft's waveguide paths soldering *Programmnye Produkty i Sistemy* **2(114)** pp 128-134.

[11] Murygin A V, Tynchenko V S, Laptenok V D, Emilova O A and Seregin Y N 2017 Modeling of thermal processes in waveguide tracts induction soldering *IOP Conference Series: Materials Science and Engineering* **173**

[12] Panteleev A V and Letova T F 2008 *Optimization methods in the examples and problems* (M.: Vyshaya Shkola)

[13] Schepetilnikov V A 1973 *The theory and design of balancing machines* (M.: Mashgiz)

A control algorithm for waveguide path induction soldering with product positioning[1]

V S Tynchenko, A V Murygin, V E Petrenko, Yu N Seregin, O A Emilova
Reshetnev Siberian State University of Science and Technology, 31, Krasnoyarsky Rabochy Av., Krasnoyarsk, 660037, Russian Federation

E-mail: vadimond@mail.ru

Abstract. In the article, the problem of developing a control algorithm is described for the induction soldering process of aluminium alloy waveguide paths. The authors suggest a solution using logic controllers in two loops: controlling the speed at which product elements heat up and controlling the waveguide assembly movement relative to the plane of the inductor. The proposed solutions are based on the analysis of the thermal processes occurring in the waveguide pipe and the flange/coupler. Based on the results of numerical experiments, forms of control actions in the system and their parameters were selected. The proposed approach to generating such a control was tested in a series of waveguide path soldering field experiments and the obtained graphs showing the heating of product elements allow the efficiency of the developed logic controllers to be confirmed. The application of the proposed approach provides the high quality regulation of the induction heating process and also the possibility to obtain reliable permanent connections between elements of waveguide paths. Through the flexible adjustment of the proposed logic controller parameters, the high versatility of their use can be demonstrated.

1. Introduction

The technological processes of induction soldering are of key importance in a number of industries. Perfection of these processes is vital in ensuring a high quality product and in improving productivity [1,2].

The theory and practice of induction heating, the calculation and design of induction plants, and the development of automatic control systems are the subjects of many works of such scientists as A. V. Sluhocki, S. E. Ryskin, V. S. Nemkov, V. D. Laptenok, A. V. Murygin S. V. Shapiro, A. S. Vasiliev, P. N. Silchenko, M. Mihnev, V. B. Demidovich, A. A. Shulyak and L. E. Roginskaya, as well as a number of Russian and foreign enterprises, such as VNIITVCH-ESTEL (G. S.-Petersburg), NKTB "Vihr'", NPP "Kuraj", NPO "Parallel" (Ufa), IJST "EKOM", (Krasnoyarsk), ABB, INDUCTOHEAT (USA). However, there are a number of topical issues requiring further research.

[1] The reported study was funded by Russian Foundation for Basic Research, Government of Krasnoyarsk Territory, Krasnoyarsk Region Science and Technology Support Fund to the research project № 16-48-242029

2. Features of induction soldering

The assembly process by means of the induction soldering of lightweight elements made of aluminium alloys is complicated by the presence of a number of external factors, the greatest difficulty in which relates to a low degree of repeatability in the non-automated, manual soldering process. These factors also concern a high level of complexity in the visual control of the heating of elements, sometimes even rendering this impossible, the distortion of the electromagnetic field of the inductor, through its interaction with various conducting bodies in close proximity to the soldering area, and the negative impact of the human factor, among others. [3,4]. The cumulative impact of these and other factors leads to uncorrectable defects in the manufacture of waveguide soldered joints. The number of defects that are unable to be corrected is of up to 5 %.

At the same time, the complex automation of the technological processes of induction soldering that should allow the influence of the above factors to be neutralized is in turn also constrained by several factors, including the insufficient study of electrothermal processes in the assembly of the induction soldering of waveguide paths, poor precision and limited capabilities of the automated control systems of induction heating processes, as well as the lack of reliable feedback systems between the automated control system unit, technological equipment and soldering area [5,6].

Since the commissioning of assembly technology for the induction soldering of waveguide paths elements, significant experience has been gained of various control systems for the automation of soldering processes (including systems using machine vision and control systems of power parameters, among others). At the same time, there are limited solutions to the challenges posed by the issues of ensuring the required accuracy of the automated control systems for soldering process, and the required characteristics of the equipment and tooling.

The technological process involved in the assembly for the induction soldering of waveguide paths can be represented as a system of interacting elements (technological operations, their cycles, methods and equipment and management systems), considered as a whole. Together, these issues have engaged scientists Faure, A., Sidorov V. I., Kuropatkin P. V., Stepanov M. F., Gladkov E. A., Reznik A. L. and others.

Since the commissioning of assembly technology for this process, significant experience has been acquired with various technological equipment, means and methods of measuring process parameters, and the process control system for automation of soldering process [7,8].

3. Automation of the induction soldering process

Methodologically, the development of automated technological systems requires a systematic approach. It involves the definition of basic input and output process parameters, the selection of control actions and the development of sensors for the main technological parameters, the identification of the control object (the construction of mathematical models of the process), and the definition of optimal modes and synthesis of the control system.

When control of the technological processes is in conjunction with mathematical models, there is a necessity to use other modern methods of identifying control objects and adaptive control system theory with feedback on key process parameters.

Authors made the decision to use contactless pyrometric sensors in developing the complex automated equipment. The main requirements for the choice was higher temperature control precision (less than 0.5%), the diameter of the controlled spot not being more than 1.5 mm. (to allow temperature control in the gap of the inductor-waveguide), electromagnetic interference immunity and its connection to a computer [9,10].

AST 250 pyrometers were selected as non-contact temperature sensors as they best satisfy the above requirements.

In the study of thermal fields during the waveguide element induction heating, it was determined that for the automation of the process of waveguide path element induction soldering 2 pyrometers are sufficicent: one for temperature control on the waveguide flange and the other on the waveguide tube.

The results of experimental studies of the induction soldering process of various sizes waveguide paths revealed the following pattern: software power control of the heating allows the reproduction of the temporal temperature characteristic only for one element of the connection waveguide tube. The temperature of the second element, the flange (coupling area), may significantly differ from the temperature of the pipe by 30÷70 degrees Celsius. With the right choice of starting distance from the flange to the inductor, it is possible to reduce the temperature difference in the vicinity of the melting solder temperature. However, it is not possible to completely eliminate the range of temperatures because of the scatter of the waveguide tube thickness, which can be over 20% of the nominal thickness. This problem can be solved by the automatic adjustment of the distance from the flange to the inductor during the technological process of soldering. The structure of the control system is double-circuit linked.

In connection with this, a new dual-circuit functional diagram for the automated process control system of induction soldering was developed (Fig. 1), by changing the power supplied to the inductor and the distance from the inductor to the heated product. Such a control scheme allows the heating of waveguide path elements to be performed according to a given law, ensuring the desired temperature distribution on the heated object.

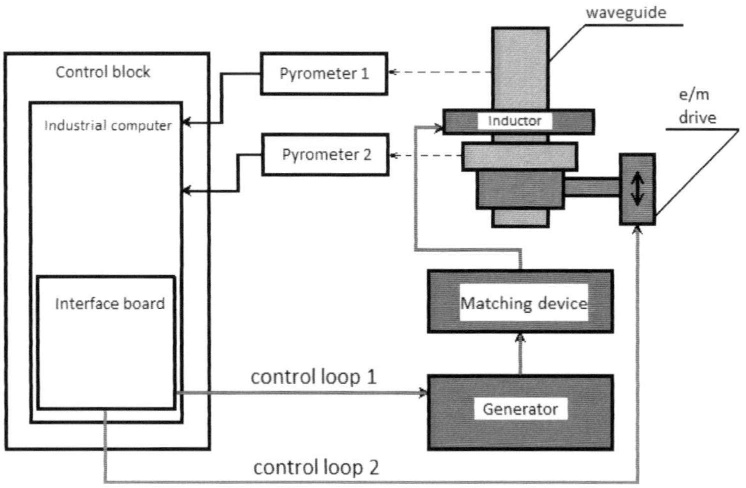

Figure 1. The structural scheme of automated equipment

Circuit 1 provides the control of power supplied to an inductor. Circuit 2 enables control of the electromechanical manipulator drive and allows the position of the soldered waveguide elements to be changed relative to the inductor.

Control of the process technological parameters is performed by the computer using error signals obtained as a result of the comparison of the program temperature with pyrometers controlling the temperature of the waveguide tube and flange [11,12].

4. Approaches to regulation in the control system of the soldering technological process

The double-loop process control system for the induction soldering of aluminium alloy waveguide paths in terms of automatic control systems can be represented by a diagram as follows (Fig. 2):

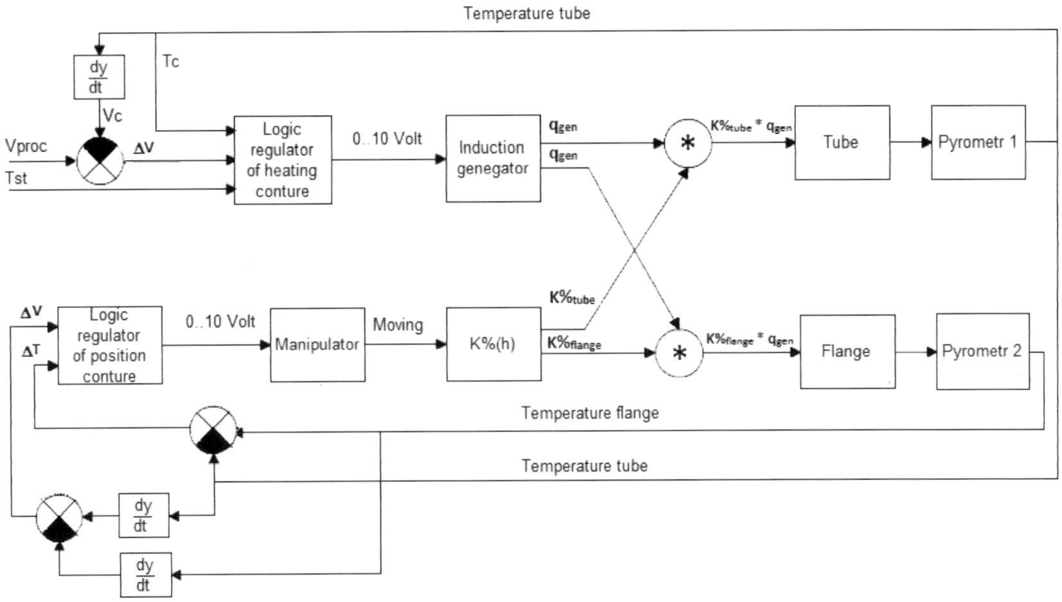

Figure 2. General scheme of the induction soldering system control

In the figure the following notation is used:

- V_{proc} is setpoint speed of heating;
- T_{st} is temperature stabilization;
- q_{gen} is generator power transferred to the brazed product;
- $K\%$, (h) is the distribution of energy between the generator elements of the product as a percentage.

For the most trivial tasks, maintaining a given heating mode is suitable for the parametric integral differential (PID) controller. However, in this task, one of the control circuits (the distance between the inductor and the device) is indirect and the use of PID control on all stages of the process is ineffective. For example, the introduction of the control loop integral component of the drive leads to a deterioration in the quality of control. It is connected to the relatively high inertia of the control object, which has already resulted in an integral link in the in the control loop, and it is not possible to improve this procedure.

Modern research considers the types and combinations of parametric (P) and differential (D) controllers, which would allow the heating rate and the temperature difference among product elements to be controlled independently of the control loop: the P-controller allows the temperature difference to be controlled, and the D-controller allows the heating rate of each of the elements to be managed.

Due to the introduction of the control loop of the workpiece position relative to the window of the inductor, the control system acquires the properties of being multi-connected and nonlinear. The mutual influence of loops can be weakened through the pulse control

IOP Conf. Series: Materials Science and Engineering **255** (2017) 012018 doi:10.1088/1757-899X/255/1/012018

system (1) so each of the circuits independently of each other are subjected to the law with constant intervals of control and waiting:

$$imp(t,\tau_1,\tau_2) = \begin{cases} 1, if \ \ mod\left(\frac{t}{\tau_1+\tau_2}\right) \le \tau_1, \\ 0, otherwise, \end{cases} \tag{1}$$

where τ_1 is the control interval; τ_2 is the waiting interval; $mod\left({}^{x_1}/{}_{x_2}\right)$ is the calculation of the remainder from dividing *x1* by *x2*; *t* is time.

Such an approach means that it is not necessary to calculate precise measures of mutual influence of loops and improves the quality of control for the following reasons:

- it eliminates the problem of high inertia of the control object relative to the control device;
- it decreases the mutual influence of the loops.

In this case, the choice of sufficiently small intervals of control and waiting (200 – 400 ms) allows renders the mutual influence of control loops negligible relative to the control.

5. The heating speed control loop

When regulating the heating speed, control is based on the logical integral controller for the deviation of the heating temperature for the soldered elements of the product above a certain threshold. In addition, separate modes of operation can be selected depending on the stage of the process:

- warm-up;
- main heating;
- stabilization phase.

As a result of exploratory research, for the first control loop (temperature control), the control dependence could be represented as a logical function (2):

$$u_1(t) = Wint(t) + \ imp(t,\tau_1,\tau_2) * \begin{cases} 0 & if\ T_c(t) < T_{sns}, \\ \frac{|\Delta V(t)|}{\Delta V(t)} * k_1, if\ T_c(t) < T_{st}\ and\ |\Delta V(t)| > V_{sns}\ and\ Tc(t) > T_{sns}, \\ \frac{|\Delta V(t)|}{\Delta V(t)} * k_2, if\ T_c(t) > T_{st}\ and\ \ |\Delta V(t)| > V_{sns}Tc(t) > T_{sns}, \\ -k_3, & if\ T_c(t) > T_{st} + T_{lim}. \end{cases} \tag{2}$$

where T_{sns} is the sensitivity temperature of the measuring device (pyrometer); $\Delta V(t)$ is the difference in the rates of increase between the controlled temperature (on one of the elements of the soldered assembly) and the heating rate program. It is calculated by the formula:

$$\Delta V(t) = V_{proc}(t) - V_c(t)$$

$V_{proc}(t)$ is the heating rate program; $V_c(t)$ is the heating rate of the soldered assemble element; V_{sns} is the tolerance threshold of the controlled heating rate; $T_c(t)$ is the temperature of the soldered assemble element (on which is the control loop); T_{st} is the stabilization mode temperature; $T_{lim}$is the permissible limit of exceeding the stabilization temperature; k_1 is the change constant of the control signal at the stage of preheating to the stabilization temperature; $k_2$is the change constant of the control signal at the temperature stabilization phase; k_3 is the change constant of the control signal when exceeding the permissible heating

IOP Conf. Series: Materials Science and Engineering **255** (2017) 012018 doi:10.1088/1757-899X/255/1/012018

limit at the stage of stabilization of the temperature; W_{int} is the integral component of the transfer function:

$$W_{int} = \frac{1}{a * s}, a \to 0.$$

6. The control loop of the workpiece position relative to the plane of the inductor

The control formation in the second control system loop is much more complex compared to the first control loop.

The control of the process of waveguide induction soldering can be divided into separate stages, each of which can be in various conditions. On the basis of system state analysis, it is possible to allocate the desired control law at a certain stage to the system state. Since the system is non-stationary, multivariable and nonlinear, the control problem cannot be solved in a trivial way using a standard control law, and requires a structural solution that covers all of the many possible states.

Table 1 presents the correlation of the technological process stages, the control object state and the ideas of control.

Table 1. Correspondence between technological process stages, states of the control object and the idea of control

Stage of the technological process	State of the control object	The idea of control
Preheating the workpiece to 300 degrees Celsius and levelling the temperature of the product elements.	Initial temperature spread of the soldered elements.	As soon as possible, eliminate the amount of misalignment.
Heating the workpiece to the soldering temperature	The transient or steady-state value in a system in which the temperatures of the elements are equal and have the same rate of increase	Ensure the convergence of the temperature of the elements and the equalization of the rates of their growth. Keeping the heating rate and the temperature difference within acceptable limits.
Melting of solder with formation of soldered joint.	A nonlinear process accompanied by a change in the temperature fields due to heat exchange between the elements	Maintaining the temperature at a given level - the melting point of the solder. Avoid overheating of the main material.

For every control idea presented in Table 1 is suitable a particular control law, which is implemented in a separate logic controller with defined coefficients. This statement is supported by the results of the mathematical modelling of the system, by searching for the main regulators using different gain values (Figures 3 and 4).

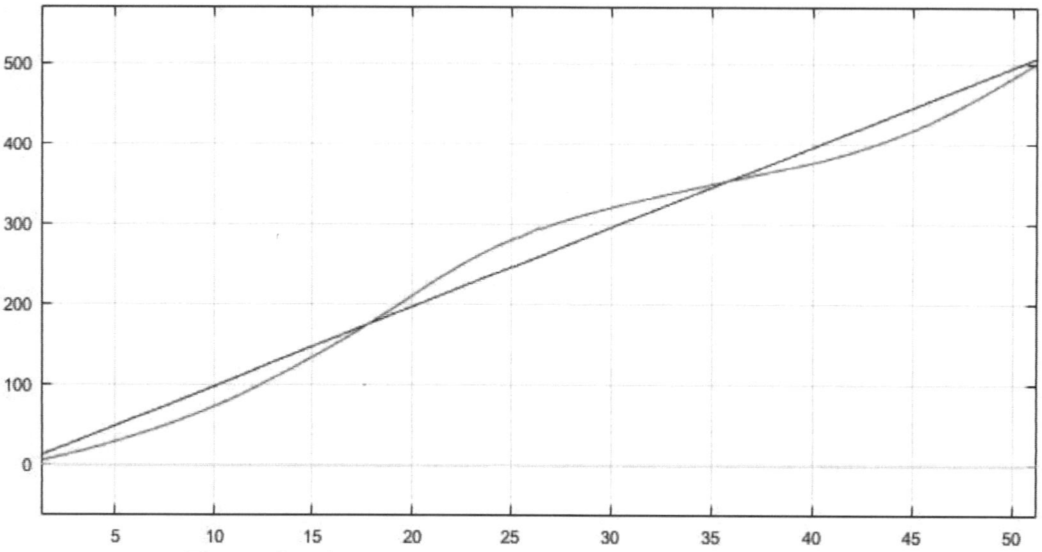

Figure 3. Simulation of the system with the P-controller.
Red graph - The temperature of the pipe, Blue graph - The temperature of the flange.
Y axis – temperature, X axis – time.

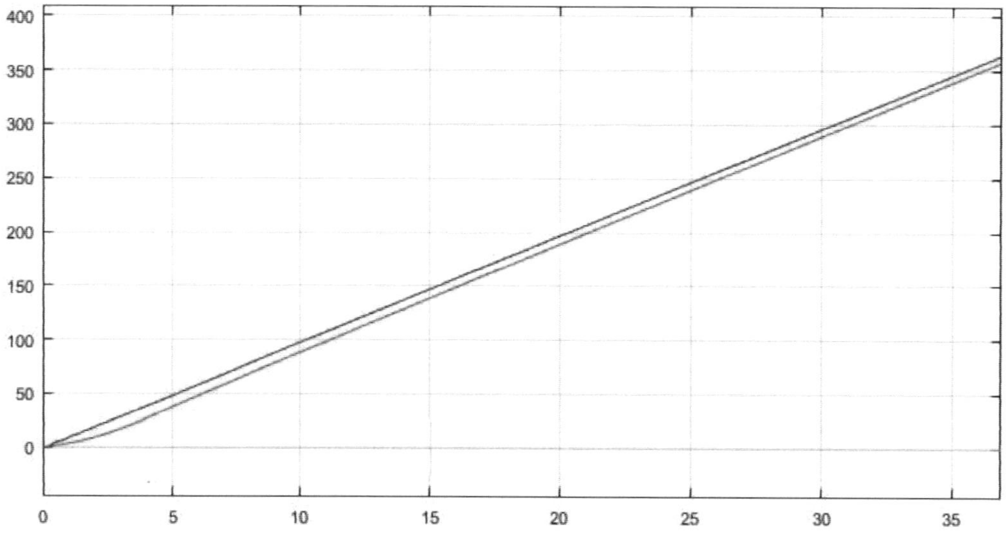

Figure 4. Simulation of the system with the D-controller.
Red graph - The temperature of the pipe, Blue graph - The temperature of the flange.
Y axis – temperature, X axis – time.

As can be seen in Figure 3, the P-controller allows graphs of the heating element temperatures to be plotted closer to each other. However, the system is in an oscillatory state, and the magnitude of the error signal is constantly changing. This P-controller with a small gain allows small deviations of the system from a given regime to be corrected.

From Figure 4 it can be seen that the D-regulator allows the rate of heating to be equalized, but the magnitude of the temperature difference between the soldered elements remains constant.

IOP Conf. Series: Materials Science and Engineering **255** (2017) 012018 doi:10.1088/1757-899X/255/1/012018

As can be deduced from the above, the development of the control law on the basis of different combinations of P- and D-controllers can fully satisfy the needs for quality control. Table 2 presents the relationship between the ideas identified for control and their implementation in the form of certain types of regulators.

Table 2. Relationship of control ideas and regulators

The idea of control	Regulator
As soon as possible, eliminate the temperature mismatch	P-regulator (with high gain).
Ensuring the convergence of temperatures and their equalization of velocities.	D-regulator. PD controller.
The retention of the heating rate and the temperature difference.	P-controller (low gain). PD - regulator.
Maintaining the temperature at the melting point of the solder	
Preventing overheating of the main material	

According to the results of experimental studies, the dependence for control for the second control circuit has been found and can be represented by a logical function (3):

$$u_2(t) = imp(t, \tau_1, \tau_2) * \begin{cases} \dfrac{|\Delta T(t)| * k_1}{k_{simple}}, & if\ |\Delta T(t)| > \Delta T_{max}\ and\ T_{tb}(t) > T_{sns}\ and\ T_{fl}(t) > T_{sns}, \\ \dfrac{(T_{fl}(t) - T_{sns}) * k_1}{k_{simple}}, & if\ T_{fl}(t) > T_{sns}\ and\ T_{tb}(t) < T_{sns}, \\ \dfrac{(T_{tb}(t) - T_{sns}) * k_1}{k_{simple}}, & if\ T_{tb}(t) > T_{sns}\ and\ T_{fl}(t) < T_{sns}, \\ \dfrac{|\Delta V(t)| * k_2}{k_{simple}}, & if\ |\Delta T(t)| < \Delta T_{max}\ and\ |\Delta V(t)| > \Delta V_{max}, \\ \dfrac{|\Delta V(t)| * k_2 - |\Delta T(t)| * k_3}{k_{simple}}, & if\ |\Delta T(t)| < \Delta T_{max}\ and\ \Delta V_{min} < |\Delta V(t)| < \Delta V_{max}, \\ \dfrac{|\Delta T(t)| * k_4}{k_{simple}}, & if\ |\Delta T(t)| < \Delta T_{max}\ and\ |\Delta V(t)| < \Delta V_{min}. \end{cases} \quad (3)$$

where T_{sns} is the sensitivity temperature of the measuring device (pyrometer); $\Delta V(t)$ is the difference in the rate of increase in temperature of elements. It is calculated as:

$$\Delta V(t) = V_{fl}(t) - V_{tb}(t);$$

$\Delta T(t)$ is the temperature difference between the assembled elements. It is calculated as:

$$\Delta T(t) = T_{fl}(t) - T_{tb}(t);$$

$T_{fl}(t)$ is the current flange temperature; $T_{tb}(t)$ is the current waveguide temperature; $V_{fl}(t)$ is the current flange heating rate; $V_{tb}(t)$ is the current tube heating rate; ΔV_{max} is the upper threshold of regulator activation; ΔV_{min} is the lower threshold of regulator activation; ΔT_{max} is the threshold for regulating the temperature of the process; $k_1 k_2, k_{3,} k_4$ are the gain factors of individual regulators; k_{simple} is the common coefficient of reduction to one order.

The above mathematical models for the regulation have already been implemented within an automated control system.

7. Experimental studies

Authors used 3 sizes of waveguide tubes and flanges/couplings when conducting experimental studies for testing the proposed approach to the control of the induction soldering of aluminium alloy waveguide paths:

1) 58×25 mm;
2) 35×15 mm;
3) 19×9.5 mm.

Figures 5, 6, 7 show the graphs of the soldering process. In each case, high-quality solder joints were obtained.

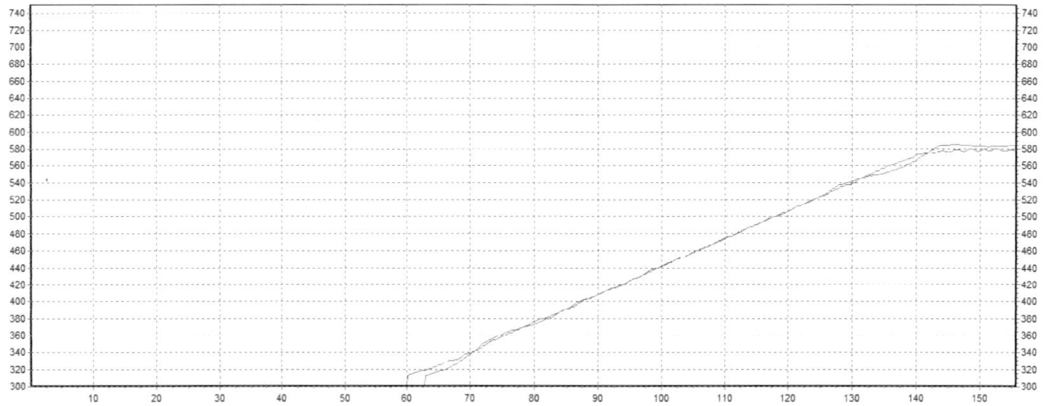

Figure 5. Soldering process for the standard size 58×25 mm
Red graph - The temperature of the pipe, Blue graph - The temperature of the flange.
Y axis – temperature, X axis – time.

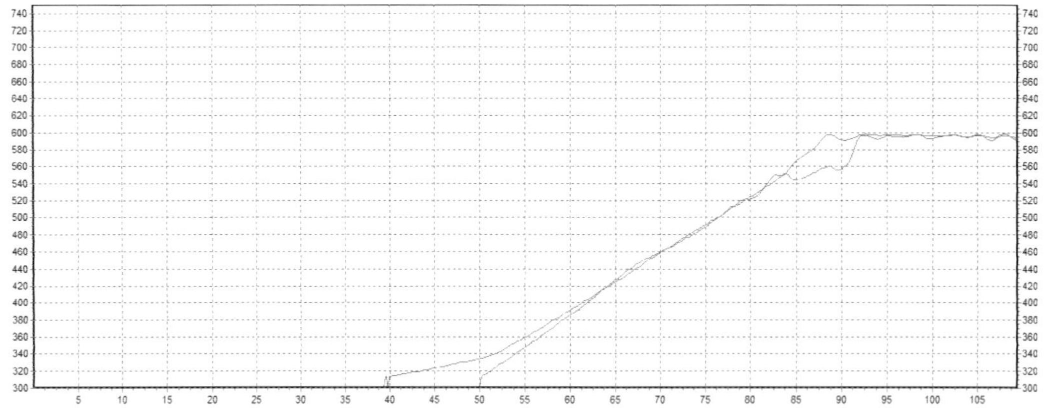

Figure 6. Soldering process for the standard size 35x15 mm
Red graph - The temperature of the pipe, Blue graph - The temperature of the flange.
Y axis – temperature, X axis – time.

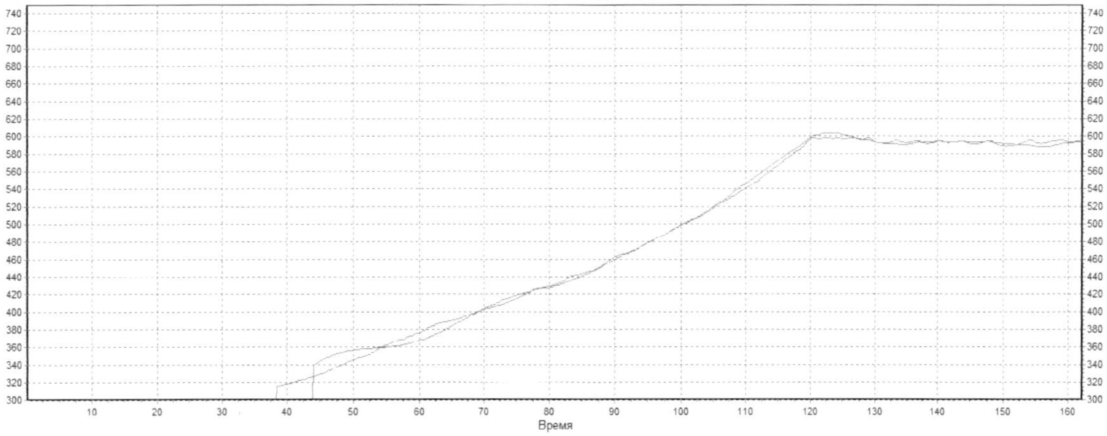

Figure 7. Soldering process for the standard size 19x9.5 mm
Red graph - The temperature of the pipe, Blue graph - The temperature of the flange.
Y axis – temperature, X axis – time.

8. Discussion of results

The graphs in Figures 5, 6 and 7 show that the differences in the sizes of soldered products do not have a significant impact on the quality of control. The proposed logic controllers (2, 3) for the induction soldering technological process can be used without reconfiguration of their settings for each individual size.

At each stage of soldering the idea of control (Tab. 1) was successfully realized, which confirms the correctness of the choice of control laws and the conditions of their application.

In Figure 5, the temperature of the product elements was recorded with a small-time difference and have a small initial misalignment, so there is almost no overshoot before melting the solder. Furthermore, upon reaching the melting temperature, the process of stabilizing the product element temperatures occurs with the subsequent formation of high quality solder joints.

In Figure 6, initial overheating of the waveguide pipe is visible, wherein the system reaches a steady state before melting the solder, after which the temperature graphs reconverge and the solder joints are formed. In addition, it is clear that the resulting non-linear process of redistribution of the product element temperature fields can be regulated by the end of the melting of the solder and the joint being completed.

In Figure, 7, as a result of the initial distribution of temperatures occurs a quickly fading oscillatory process, after which the system in a steady state brings the technological process to the stage of the solder melting and its successful completion.

In all these cases, high-quality solder joints were achieved. In addition, all of the graphs showing the processes of induction soldering (Fig. 5, 6 and 7) show that the choice of pulse character for controlling the automated system reduces the impact of the cross-coupling of loops to an insignificant level and improves the quality of control.

9. Conclusion

The following results were obtained in the current study:

1) the form and the types of regulators that allow the induction soldering process for aluminium alloy waveguide paths to be controlled without any need to reconfigure their internal state, which indicates the versatility of their use for a wide range of waveguide sizes;

2) the application of impulse control in the system allowed the cross-influence in the system to be reduced to a negligible level, thereby providing an appropriate regulatory process;

3) using the developed control system allows a high-quality soldering connection to be obtained for aluminium alloy waveguide paths.

Thus, the study addresses the problem of generating effective control for the double-circuit system for the induction soldering of aluminium alloy waveguide paths, which assures the high quality of soldered joints and reduces the human impact on the technological process.

References

[1] Zlobin S K, Mikhnev M M, Laptenok V D and Zaitsev R V 2012 Features of soldering waveguide distribution paths of aluminum alloys using induction heating source *Reshetnev reading : proceedings of the XVI Intern. scientific. conf.* **vol 2** pp 192-193

[2] Silchenko P N, Ankudinov A V and Kudryavtsev I V 2012 Ensuring the strength and accuracy of large-sized waveguide-distributive systems of space communication devices *Problemy mashinostroeniya i nadezhnosti mashyn* **№1** pp 112-117

[3] Vologdin V V, Kushch E V and Assam V V 1989 *Induction soldering* (L.: Mechanical engineering)

[4] Kudryavtsev I V., Barykin E S and Gotselyuk O B 2013 Mathematical model of heating of the waveguide during transmission of high signal power *Young scientist* **№9** pp 52-57

[5] Laptenok V D, Zlobin S K, Mihnev M M and Bocharov A N 2013 Features of the production of the waveguide-junction of paths of antenna-feeder devices of spacecraft *Vestnik SibGAU* vol **6(52)** pp 196-201

[6] Laptenok V D, Dolgopolov B B and Bocharov A N 2009 The technique of debugging of technological process of soldering the waveguide paths (Krasnoyarsk : SibGAU) 112 p.

[7] Slugocki A E and Ryskin S E 1974 *Inductors for induction heating* (Leningrad : Energy)

[8] Slugocki A E 1989 *Inductors* (L.: Engineering)

[9] Tynchenko V S, Bocharov A N, Laptenok V D, Seregin Yu N and Zlobin S K 2016 The software of technological process of spacecraft's waveguide paths soldering *Programmnye Produkty i Sistemy* **№2(114)** pp 128-134

[10] Tynchenko V S, Murygin A V, Emilova O A, Bocharov A N and Laptenok V D 2016 The automated system for technological process of spacecraft's waveguide paths soldering *IOP Conference Series: Materials Science and Engineering* **155**

[11] Murygin A V, Tynchenko V S, Laptenok V D, Emilova O A and Bocharov A N 2017 Complex of automated equipment and technologies for waveguides soldering using induction heating *IOP Conference Series: Materials Science and Engineering* **173**

[12] Murygin A V, Tynchenko V S, Laptenok V D, Emilova O A and Seregin Y N 2017 Modeling of thermal processes in waveguide tracts induction soldering *IOP Conference Series: Materials Science and Engineering* **173**

IOP Conf. Series: Materials Science and Engineering **255** (2017) 012019 doi:10.1088/1757-899X/255/1/012019

Reflection electron energy loss spectroscopy of structures based on silicon and transition metals

A S Parshin[1], A Yu Igumenov[1], Yu L Mikhlin[2], O P Pchelyakov[3,1], V S Zhigalov[4,1]

[1] Reshetnev Siberian State University of Science and Technology, 31 Krasnoyarskiy Rabochiy av., Krasnoyarsk, 660037, Russia

[2]Institute of Chemistry and Chemical Technologies of the Siberian Branch of the Russian Academy of Sciences, 50/24 "Akademgorodok", Krasnoyarsk, 660036, Russia

[3]Rzhanov Institute of Semiconductor Physics of the Siberian Branch of the Russian Academy of Sciences, 13 "Academician Lavrentjev Av.", Novosibirsk, 630090, Russia

[4]Kirensky Institute of Physics of the Siberian Branch of the Russian Academy of Sciences, 50/38 "Akademgorodok", Krasnoyarsk, 660036, Russia

E-mail: aparshin2010@mail.ru

Abstract. The investigation of iron silicides $FeSi_2$, FeSi and Fe_5Si_3 with the methods of reflection electron energy loss spectroscopy and inelastic electron scattering cross-section spectroscopy was carried out. It is shown that the shape and peak energy position of electron energy loss spectra are almost identical to silicides with different composition, while the amplitude of inelastic electron scattering cross-section spectra decreases with increasing of iron content. The decomposition of inelastic electron scattering cross-section spectra of $FeSi_2$, FeSi and Fe_5Si_3 to Tougaard peaks is used for unresolved energy loss peaks analysis, determination its energies and identification bulk and surface peaks. The amplitude of fitting peak describing bulk plasmon excitation can be used for identification of the iron silicides with different composition.

1. Introduction

Fe-Si system is promising for basic research and applied applications in devices of nanoelectronics, spintronics and photonics [1]. Elemental analysis of ferrosilicon structures is complicated by the formation of the silicides because the chemical shifts in X-ray photoelectron spectra (XPS) practically do not depend on the iron silicide composition [1, 2] and bulk plasmon energy in reflection electron energy loss spectra (REELS), obtained by different authors give conflicting results [3].

In this paper the results of iron silicides investigation with the methods of REELS and inelastic electron scattering cross-section spectroscopy are presented. It is shown that the amplitude of the inelastic electron scattering cross-section spectra is more sensitive to the elemental composition of iron silicides that REELS peaks energies. The decomposition of experimental inelastic electron scattering cross-section spectra to the Tougaard peaks is used for unresolved loss peaks analysis, determination its energies and identification bulk and surface peaks. The amplitude of fitting peak describing bulk plasmon excitation can be used for identification of the iron silicides with different composition.

2. Experimental

Iron silicides $FeSi_2$, FeSi, Fe_5Si_3 produced by melting a mixture of iron and silicon in atomic ratios of 1×2, 1×1, 5×3 under high vacuum conditions using the vacuum deposition installation UVN-2M-1. The mixture was maintained at the melting temperature for 15 min, after which the annealed alloy was crushed and again melted for 15 min. The 1 mm thick washers were cut of bulk samples. Before the spectroscopic investigations the washers were polished.

The spectroscopic experiments were carried out using a photoelectron spectrometer SPECS (German production) equipped with the spherical energy analyzer PHOIBOS MCD9, double anode X-ray tube as an X-ray source, and a Microfocus EK-12-M electron gun (STAIB Instruments) for excitation of the REEL spectra. The surface contaminations, protecting layers and oxide layers were removed applying Ar^+ ion etching (accelerating voltage of 2.5 kV, ion current of 15 μA) with an ion source IQE-12/38 (SPECS) in the analytical chamber of the spectrometer before the REELS measurements; the completeness of the procedure was controlled via the relevant photoelectron and Auger spectra.

REEL spectra were acquired in the interval of 150 eV below the elastic peak with an energy step of 0.1 eV. Energy loss T was calculated as the difference between the primary electron beam energy E_0 (zero loss) and the reflected electron energy E, $T = E_0 - E$. The primary electron beam energies were 300, 600, 1200, 1900, 3000 eV, and the full width at half-maximum (FWHM) of the primary electron beam was less than 1 eV.

The inelastic electron scattering cross section spectra (so-called $K\lambda$-spectra [4]) which are the products of inelastic mean free path λ and inelastic scattering cross-section $K(E_0, T)$ were calculated from the experimental reflection electron energy loss spectra with the software package QUASES™ XS REELS (Quantitative Analysis of Surfaces by Electron Spectroscopy cross section determined by REELS) [6] based to the algorithm suggested in [7]. Inelastic electron scattering cross section spectra indicate the probability of electron energy loss T in a single scattering event. $K\lambda$-spectra can be used for quantitative determination of element concentrations in two-component composite structures. The ability of quantitative determination of atomic element concentrations in Fe_xSi_{1-x} [8, 9], Mn_xSi_{1-x} [7] and Ge_xSi_{1-x} [10] systems were investigated earlier with the dependence of the maximum value of $K\lambda$-spectra for the standard samples.

3. Results and discussion

3.1. The comparative analysis of the reflection electron energy loss spectra and the inelastic electron scattering cross-section spectra

Fig. 1 (a) shows REELS for elemental Si, Fe and silicides $FeSi_2$, FeSi, Fe_5Si_3, obtained at the primary electron energy 1200 eV. The main peak energy for Si is 17.1 ± 0.2 eV, for Fe - 22.8 ± 0.4 eV, for $FeSi_2$, FeSi, Fe_5Si_3 – respectively, 20.8 ± 0.1, 21.0 ± 0.1, 21.5 ± 0.1 eV. These values are close to the energies of bulk plasmons of these materials [3]. These energies defined using the spectra in differential form, since this form is more accurate and allows determining the energy of the unresolved bulk plasmon in spectra of iron. On the spectra of silicon and iron silicides multiple energy loss due to excitation of bulk plasmons are present. At energy loss of 10.7 ± 0.3 eV in the spectrum of silicon the feature corresponding to the excitation of surface plasmon [3] is observed. In the spectrum of iron intense losses at the low-energy region and the one-electron transition M_{23} with energy losses 54.3 ± 0.3 eV [11-14] are observed.

As can be seen from the Fig. 1 (a), REELS spectra of iron silicides $FeSi_2$, FeSi, Fe_5Si_3 slightly different in bulk plasmon energy and are almost identical in shape and intensity, which is measured in arbitrary units. The surface plasmon is unresolved in the spectra of iron silicides. Elemental Si and Fe stand out: silicon has a much lower energy of bulk plasmon and iron has a more extended form, due to intense energy loss for surface excitation.

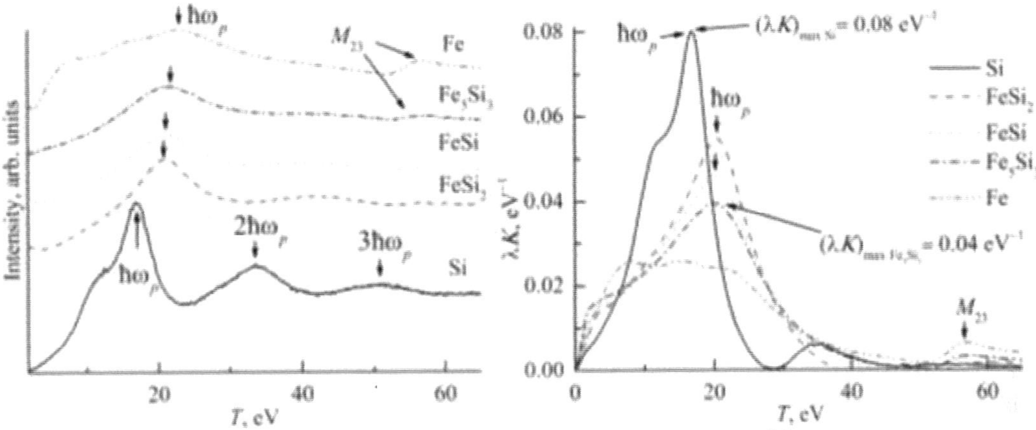

a) b)

Figure 1. Reflection electron energy loss spectra (a) and inelastic electron scattering cross-section spectra (b)

At Fig. 1 (b) the comparison of the inelastic electron scattering cross-section spectra for silicides $FeSi_2$, FeSi, Fe_5Si_3 and elemental Si, Fe, obtained with primary electron energy 1200 eV, is carried out. Differences in $K\lambda$-spectra of these materials are more significant that in the REELS. First, the spectrum amplitude decreases with increasing of Fe content. Secondly, the spectrum width increases with increasing of Fe content. The main peak energy in REELS is almost independent of primary electron energy and increases monotonically with increasing of primary electron energy in inelastic electron scattering cross-section spectra. With increasing of primary electron energy increases the depth analysis, resulting in decreasing of surface-like excitation intensity and increasing of bulk-like excitation intensity. This intensity redistribution causes a change of whole spectrum shape and peak position. This assumption is confirmed by simulation of $K\lambda$-spectrum as a superposition of two unresolved peaks [15]. The presence of unresolved surface plasmon in $K\lambda$-spectra of iron silicides manifests in decreasing of FWHM.

In Table 1 the bulk plasmon energies obtained using experimental REELS and $K\lambda$-spectra in differential form are shown.

Table 1. The bulk plasmon energies (eV)

	REELS	$K\lambda$-spectra	Fitting $K\lambda$-spectra
Silicide	The bulk plasmon energy, eV		
$FeSi_2$	20.8 ± 0.1	20.6 ± 0.2	21.1 ± 0.0
FeSi	21.0 ± 0.1	20.0 ± 0.6	20.9 ± 0.1
Fe_5Si_3	21.5 ± 0.1	20.0 ± 0.7	20.7 ± 0.0

Thus, firstly, the peaks positions in inelastic electron scattering cross-section spectra depend on the different processes relative contributions and may differ from the true energy loss peaks. In determining the energy peaks in *Kλ*-spectra the obtained values can be influenced by surface excitations. Second, the composition of the iron silicides can not be reliably determined from the bulk plasmon energy in REELS and Kλ-spectra, because silicides of different composition has an overlap of bulk plasmon energy ranges.

3.2. The fine structure of inelastic electron scattering cross-section spectra

The variation of inelastic electron scattering cross section spectrum shape with the primary electron energy is affected by the variation of surface-like and volume-like excitation intensities. In Ref. [9, 10] the method of inelastic electron scattering cross-section spectra fine structure analysis was suggested. The fine structure analysis is based on fitting *Kλ*-spectra to the 3 parameters Lorentzian-type formula of Tougaard [11]:

$$\lambda K = \frac{BT}{\left(C - T^2\right)^2 + DT^2}.$$

B, C, D are fitting parameters and have special values for different elements [11]. The peak intensity depends on parameter *B*, peak energy depends on parameter *C*, peak width and, indirectly, peak energy depend on parameter *D*.

We chose the 3 parameters Lorentzian-type formula of Tougaard because it describes the inelastic electron scattering mechanisms. The *Kλ*-spectra fitting to the Tougaard peaks can be used to analyze different mechanisms of inelastic electron scattering such as volume plasmon excitations, surface plasmon excitations and interband transitions.

The results of fitting the $FeSi_2$ and Fe_5Si_3 *Kλ*-spectra that were obtained with primary electron energies 600 and 1900 eV are shown in figures 2 and 3. The decomposition of inelastic electron scattering cross-section spectra for elemental Fe, Si and Mn were carried out in [15, 19, 20, 21]. The decomposition of inelastic electron scattering cross-section spectra for silicides $FeSi_2$ and FeSi were carried out in [22, 23].

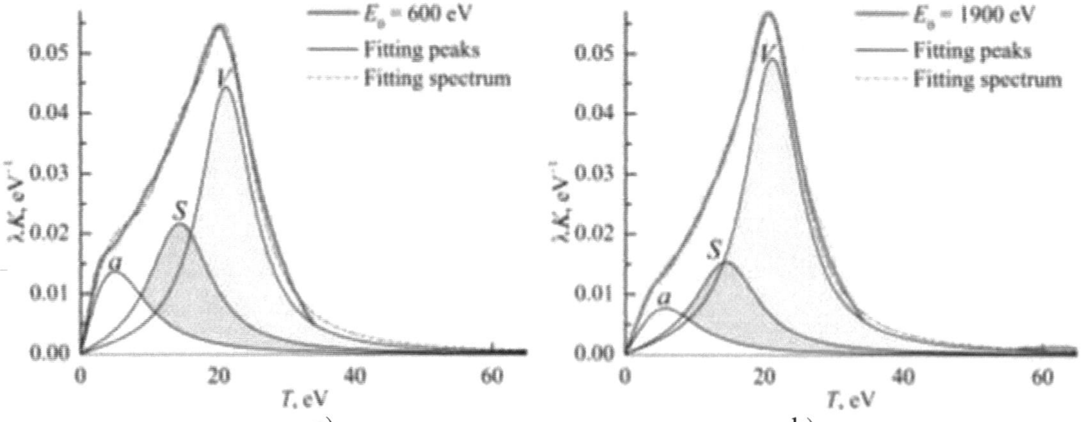

Figure 2. The fine structure of inelastic electron scattering cross-section spectra for $FeSi_2$ obtained with primary electron energies 300 (a) and 1900 (b) eV

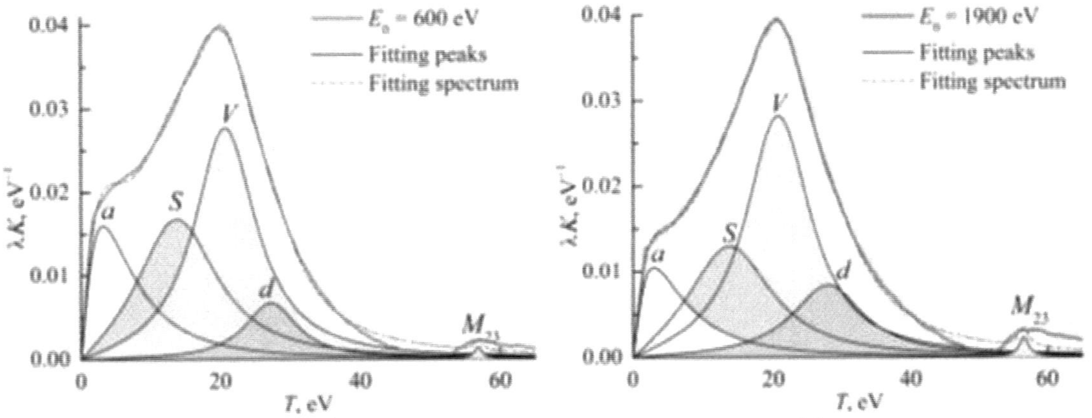

Figure 3. The fine structure of inelastic electron scattering cross-section spectra for Fe_5Si_3 obtained with primary electron energies 300 (a) and 1900 (b) eV.

Peak *V* corresponds to volume plasmon. This peak has gradually increasing primary energy dependence of amplitude which indicate its volume origin (Fig. 5). The peak *S* corresponds to surface plasmon. This peak has opposite to peak *V* gradually decreasing primary energy dependence of amplitude. Surface plasmon is unresolved in experimental spectra for silicides and first its energy sets smaller by a factor of $\sqrt{2}$ to bulk plasmon energy (classical plasmon model [24]) then surface plasmon energy and parameters of other peaks were varied for the best least squared fit to the experimental spectra. This peak was unresolved in REELS and *Kλ*-spectra for silicides but the method of fine structure analysis allows determining its energy, intensity and origin. The M_{23} peak related to the electron transition from the level M_{23} and its energy loss is close to the literary data for elemental Fe. Peaks *a* and *d* appearingly related to the interband transitions.

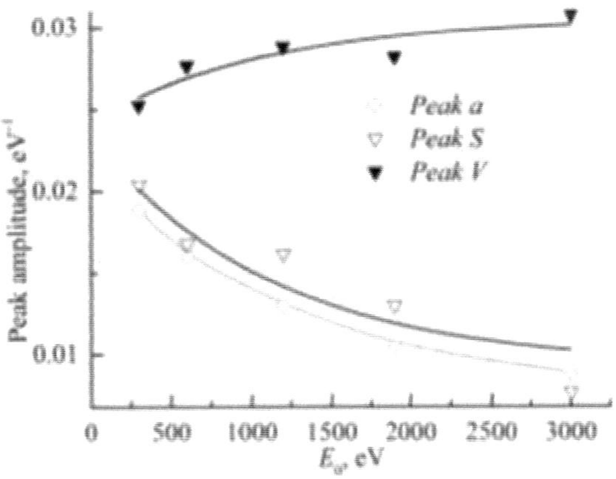

Figure 4. The primary energy dependences of the fitting peaks amplitude

The fine structure analysis method allows determining the peaks energies more accurately than using experimental spectra. Firstly, this method balancing out an effect of summation of different losses intensities to the energies of spectrum maximums. Secondly, the analysis of unresolved peaks becomes possible. Third, the definition of fitting peak energy is technically more authentic than for the experimental one.

Fig. 5 shows the primary energy dependence of the peak amplitude describing the energy loss due to excitation of the bulk plasmon A_V. For all the investigated materials peak V amplitude is varied enough to be able to uniquely identify Si, Fe, $FeSi_2$, FeSi, Fe_5Si_3 using this parameter.

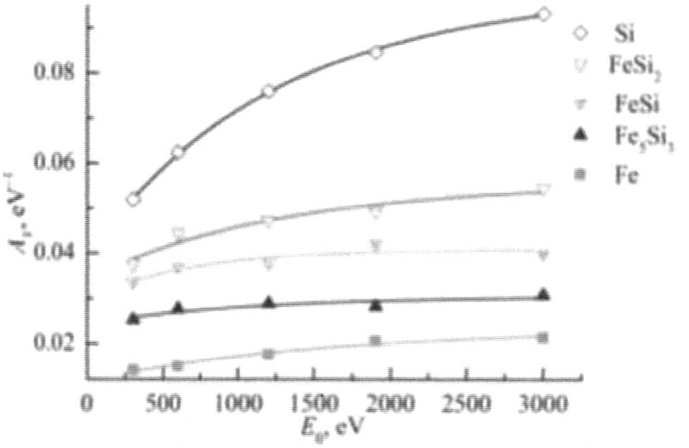

Figure 5. The primary energy dependences of the fitting peak V amplitude for different materials

If the energy peaks positions in experimental REELS and *Kλ*-spectra can not be used to distinguish silicides with different composition, the fitting peak *V* amplitude is can be used to identify silicides with different composition.

4. Conclusions

The reflection electron energy loss spectra and the inelastic electron scattering cross-section spectra for silicides $FeSi_2$, FeSi, Fe_5Si_3 and elemental Si, Fe were investigated. It is shown that the composition of the iron silicides can not be reliably determined using the peak energies in REELS and *Kλ*-spectra. If REELS of silicides with different composition are substantially identical in shape and main peak energy, the amplitude of *Kλ*-spectra noticeably decreases with increasing of Fe content.

Decomposition of $FeSi_2$, FeSi, Fe_5Si_3 *Kλ*-spectra to Tougaard peaks allowed to identify the unresolved loss peaks, determine its energies, intensities and origin. It is shown that the peak amplitude describing the energy loss due to excitation of the bulk plasmon can be used to identify silicides with different composition.

The reported study was funded by RFBR and Krasnoyarsk Krai according to the research project № 17-42-240884.

References

[1] Ohtsu N, Oku M, Satoh K, Wagatsuma K 2013 *Appl. Surf. Sci.* **264** pp 219-224
[2] Ohtsu N, Oku M, Nomura A, Sugawara T, Shishido T, Wagatsuma K 2008 *Appl. Surf. Sci.* **254** pp 3288-3294
[3] Lifshits V G, Lunyakov Yu V 2004 Electron Energy Loss Spectra of Surface Phases on Silicon *Vladivostok: Dal'nauka.*
[4] Gergely G 2002 *Prog. Surf. Sci.* **71** pp 31-88

[5] Tougaard S QUASES – Software packages to characterize surface nano-structures by analysis of electron spectra http, available at: http://www.quases.com/.
[6] Tougaard S, Chorkendorff I 1987 *Phys Rev. B* **35** pp 6570-6577
[7] Parshin A S, Pchelyakov O P, Dolbak A E, Ol'shanetskii B Z 2013 *Journal of Surface Investigation. X-ray, Synchrotron and Neutron Techniques* **7** pp 505-508
[8] Parshin A S, Aleksandrova G A, Dolbak A E, Pchelyakov O P, Ol'shanetskii B Z, Ovchinnikov S G, Kushchenkov S A 2008 *Tech. Phys. Letters* **34** pp 381-383
[9] Parshin A S, Kushchenkov S A, Aleksandrova G A, Ovchinnikov S G 2011 *Tech. Phys.* **56** pp 656-661
[10] Parshin A S, P'yanovskaya E P, Pchelyakov O P, Mikhlin Yu L, Nikiforov A I, Timofeev V A, Esin M Yu 2014 *Semiconductors* **48** pp 224-227
[11] Egert B, Panzner G, 1981 *J. Phys. F* **11** pp L233- L239
[12] Egert B, Panzner G 1984 *Phys. Rev. B* **29**, **4** pp 2091-2101
[13] Sakisaka Y, Myano T, Onchi M 1984 *Phys. Rev. B* **30**, **12** pp 6849-6855
[14] Feldman L C, Mayer J W 1986 Fundamentals of Surface and Thin Film Analysis *New York: Elsevier Science*
[15] Parshin A S, Igumenov A Yu, Mikhlin Yu L, Pchelyakov O P, Zhigalov V S 2016 *Physics of the Solid State* **58** pp 908-914
[16] Tougaard S 1997 *Surf. Interface Anal.* **25** pp 137-155
[17] Gergely G, Menyhard M, Gurban S, Sulyok A, Toth J, Varga D, Tougaard S 2001 *Solid State Ion.* **141–142** pp 47-51
[18] Orosz G T, Gergely G, Gurban S, Menyhard M, Toth J, Varga D, Tougaard S 2003 *Vacuum* **71** pp 147-152
[19] Parshin A S, Igumenov A Yu, Mikhlin Yu L, Pchelyakov O P, Nikiforov A I, Timofeev V A 2015 *Semiconductors* **49**, **4** pp 423-427
[20] Parshin A S, Igumenov A Yu, Mikhlin Yu L, Pchelyakov O P, Nikiforov A I, Timofeev V A 2014 *Vestn. Sib. Gos. Aerokosm. Univ.* **4 (56)** pp 230-235
[21] Parshin A S, Igumenov A Yu, Mikhlin Yu L, Pchelyakov O P, Zhigalov V S 2016 *IOP Conference Series: Materials Science and Engineering* **122** pp 012025, 1-7
[22] Parshin A S, Igumenov A Yu, Mikhlin Yu L, Pchelyakov O P, Zhigalov V S, 2016 *Technical Physic* **61 (9)** pp 1418-1422
[23] Parshin A S, Igumenov A Yu, Mikhlin Yu L, Pchelyakov O P, Zhigalov V S, 2016 *Izvestiya Vysshikh Uchebnykh Zavedenii, Fizika* **59, 10** pp 82-86
[24] Raether H 1988 Surface Plasmons on Smooth and Rough Surfaces and on Gratings *Heidelberg: Springer-Verlag Berlin* 135 p

IOP Conf. Series: Materials Science and Engineering **255** (2017) 012020 doi:10.1088/1757-899X/255/1/012020

8-order filter based on 2-D photonic crystal with dual-mode microstrip resonators

B A Belyaev[1,2], **S A Khodenkov**[1]
[1]Department of Physics, Reshetnev Siberian State University of Science and Technology, Krasnoyarsk, 660037, Russia
[2]Laboratory of electrodynamics and microwave electronics, Kirensky Institute of Physics, Krasnoyarsk, 660036, Russia

E-mail: has-sibgau@mail.ru

Abstract. The design of 8-order band-pass filter based on 2-D photonic crystal is suggested. Strip conductors in all dual-mode resonators are in the form of «irregular» rectangular frame with a gap and are connected at certain points with the ground. At optimal placement of such resonators on a substrate attenuation pole is observed on the amplitude-frequency characteristic of the filter, near low and high bandwidth slope that significantly improves its frequency-selective properties. The change of gap between input and output resonators allows to control the frequency of these poles.

1. Introduction

It is known that photonic crystals are artificial periodic structures, the irregularities in the dimensions of which are comparable to the wavelength of the electromagnetic wave [1, 2]. Currently, on the basis of one-dimensional and two-dimensional (2-D) crystals various perspective frequency-selective devices, including microwave ones are developed and studied [3, 4].

This paper presents the investigation results of microstrip band-pass filter of a new design constructed on the basis of two-dimensional photonic crystal. Its amplitude-frequency characteristics calculated by means of electrodynamic numerical analysis of 3-D models are in good agreement with the experiment, in this regard theoretical study of such 2-D structures can be considered reliable. Filter tune is performed by «manual» parametric synthesis, for this reason the substrate with high dielectric constant $\varepsilon = 80$, $h = 1$ mm thickness (material – TBNS ceramics) was used.

2. Modelling results

The topology of strip conductors of the proposed microstrip filter is shown in figure 1(a). At the same time four dual-mode resonators *I* - *IV* of a structure represent two-dimensional (with inhomogeneity along the *x*-axis and along the *y*-axis) microstrip photonic crystal structure. The strip conductor of each resonator is in form of «irregular» rectangular frame with a gap and has a connection with the ground near the centre (see figure 1(a)). Such grounding can be realized by through hole in the dielectric substrate of 0.2×0.2 mm^2 filled with conductive material. So, in fact, each of these dual-mode resonators of band-pass filter is a pair of connected quarter-wave resonators that allows to miniaturize frequency-selective device significantly.

As it has been shown by the example of 2-D structure with single-mode irregular microstrip resonators [5], there are several options for the location of resonators on a substrate, in this connection, as it should be expected, filters have significantly different frequency-selective properties. Therefore, having optimum clearances between the strip conductors, the signal is transmitted sequentially from the input *I* resonator to the *II*, *III* and finally to the output *IV* resonator.

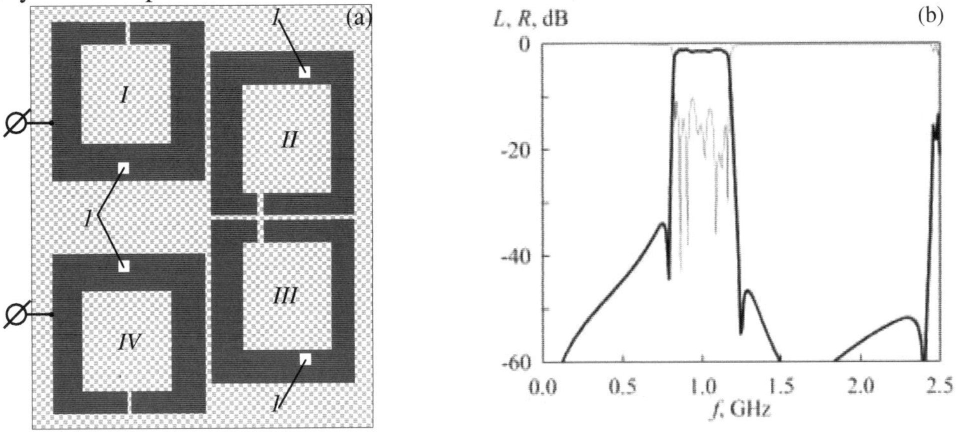

Figure 1. (a) The topology of strip conductors of 8-order filter; **(b)** Amplitude-frequency characteristics of 8-order filter. *1* – conductors' connection points with the ground

As it is shown in figure 1(b), on the amplitude-frequency characteristics of the filter near the low and high slopes of bandwidth the attenuation pole is observed that increases the steepness of the slopes greatly. Minimum power loss in the bandwidth does not exceed -1.1 dB at its relative width of ~ 34%.

It is worth noting that increasing of clearance between the conductors of *I* and *IV* resonators can simultaneously increase power suppression both in low-frequency and high frequency stopbands, but it is accompanied by the decrease of steepness of passband both slopes.

Therefore, the proposed microstrip 8-order band-pass filter based on 2-D photonic crystal has high frequency-selective properties.

Acknowledgments

This study was supported by the Ministry of Education and Science of the Russian Federation, grant MK-9119.2016.8 of the President of the Russian Federation for State Support of Young Russian Scientists.

References

[1] Escorcia-Garcıa J and Mora-Ramos M E 2009 *PIERS Online* **5** 36
[2] Belyaev B A, Voloshin A S and Shabanov V F 2005 *Dokl. Phys.* **50** 337
[3] Wu X, Yamilov A, Liu X, Li S, Dravid V P, Chang R P H and Cao H 2004 *Appl. Phys. Lett.* **85** 3657
[4] Passos M G, Fernandes H C C and da F. Silva P H 2007 *PIERS Online.* **3** 695
[5] Belyaev B A and Khodenkov S A 2016 13th International Scientific-Technical Conference on Actual Problems of Electronic Instrument Engineering vol 1 (Novosibirsk: Novosibirsk State Technical University) p 146